Springer-Textbook

Prof. Dietmar Gross
received his Engineering Diploma in Applied Mechanics and his Doctor of Engineering degree at the University of Rostock. He was Research Associate at the University of Stuttgart and since 1976 he is Professor of Mechanics at the University of Darmstadt. His research interests are mainly focused on modern solid mechanics on the macro and micro scale, including advanced materials.

Prof. Werner Hauger
studied Applied Mathematics and Mechanics at the University of Karlsruhe and received his Ph.D. in Theoretical and Applied Mechanics from Northwestern University in Evanston. He worked in industry for several years, was a Professor at the Helmut-Schmidt-University in Hamburg and went to the University of Darmstadt in 1978. His research interests are, among others, theory of stability, dynamic plasticity and biomechanics.

Prof. Jörg Schröder
studied Civil Engineering, received his doctoral degree at the University of Hannover and habilitated at the University of Stuttgart. He was Professor of Mechanics at the University of Darmstadt and went to the University of Duisburg-Essen in 2001. His fields of research are theoretical and computer-oriented continuum mechanics, modeling of functional materials as well as the further development of the finite element method.

Prof. Wolfgang A. Wall
studied Civil Engineering at Innsbruck University and received his doctoral degree from the University of Stuttgart. Since 2003 he is Professor of Mechanics at the TU München and Head of the Institute for Computational Mechanics. His research interests cover broad fields in computational mechanics, including both solid and fluid mechanics. His recent focus is on multiphysics and multiscale problems as well as computational biomechanics.

Prof. Sanjay Govindjee
received his SB from MIT and an MS and PhD from Stanford University in mechanical engineering. He was an engineering analyst at the Lawrence Livermore National Laboratory (1991–93) and Professor of Mechanics at ETH Zurich (2006–08). Currently he is a Chancellor's Professor and Professor of Civil Engineering at the University of California, Berkeley (1993–2006, 2008–present). His expertise lies in computational mechanics and the modeling of materials.

Dietmar Gross · Werner Hauger
Jörg Schröder · Wolfgang A. Wall
Sanjay Govindjee

Engineering Mechanics 3

Dynamics

2nd Edition

 Springer

Prof. Dr. Dietmar Gross
TU Darmstadt
Division of Solid Mechanics
Franziska-Braun-Str. 7
64287 Darmstadt
Germany
gross@mechanik.tu-darmstadt.de

Prof. Dr. Jörg Schröder
Universität Duisburg-Essen
Institute of Mechanics
Universitätsstr. 15
45141 Essen
Germany
j.schroeder@uni-due.de

Prof. Dr. Werner Hauger
TU Darmstadt
Division of Solid Mechanics
Franziska-Braun-Str. 7
64287 Darmstadt
Germany
hauger@mechanik.tu-darmstadt.de

Prof. Dr. Wolfgang A. Wall
TU München
Computational Mechanics
Boltzmannstr. 15
85747 Garching
wall@lnm.mw.tum.de

Prof. Sanjay Govindjee
University of California
Structural Engineering
Mechanics and Materials
779 Davis Hall
Berkeley California, 94720-1710
USA
s_g@berkeley.edu

This is the second edition and the first edition was ISBN 978-3-642-14018-1, published by Springer in 2011.

ISBN 978-3-642-53711-0   ISBN 978-3-642-53712-7 (eBook)
DOI 10.1007/978-3-642-53712-7
Springer Heidelberg Dordrecht London New York

Library of Congress Control Number: 2013958132

Printed on acid-free paper

Springer is part of Springer Science+Business Media (www.springer.com)

Preface

Dynamics is the third volume of a three-volume textbook on Engineering Mechanics. Volume 1 deals with *Statics*; Volume 2 contains *Mechanics of Materials*. The original German version of this series is the bestselling textbook on Engineering Mechanics in German speaking countries; its 12th edition has just been published.

It is our intention to present to engineering students the basic concepts and principles of mechanics in the clearest and simplest form possible. A major objective of this book is to help the students to develop problem solving skills in a systematic manner.

The book developed out of many years of teaching experience gained by the authors while giving courses on engineering mechanics to students of mechanical, civil and electrical engineering. The contents of the book correspond to the topics normally covered in courses on basic engineering mechanics at universities and colleges. The theory is presented in as simple a form as the subject allows without being imprecise. This approach makes the text accessible to students from different disciplines and allows for their different educational backgrounds. Another aim of the book is to provide students as well as practising engineers with a solid foundation to help them bridge the gaps between undergraduate studies, advanced courses on mechanics and practical engineering problems.

A thorough understanding of the theory cannot be acquired by merely studying textbooks. The application of the seemingly simple theory to actual engineering problems can be mastered only if the student takes an active part in solving the numerous examples in this book. It is recommended that the reader tries to solve the problems independently without resorting to the given solutions. In order to focus on the fundamental aspects of how the theory is applied, we deliberately placed no emphasis on numerical solutions and numerical results.

We gratefully acknowledge the support of Dr.-Ing. Vera Ebbing and the cooperation of the staff of Springer who were responsive to our wishes and helped to create the present layout of the books.

Darmstadt, Essen, München, and Berkeley, Spring 2014

D. Gross

W. Hauger

J. Schröder

W.A. Wall

S. Govindjee

Table of Contents

Introduction

The primary task of mechanics is the description and prediction of the motion of bodies along with the associated forces. The subject of mechanics can be broken into the disciplines of *statics* and *dynamics*. The subject of statics is the study of bodies in equilibrium. Dynamics, on the other hand, deals with bodies in motion. It is further sub-divided into the subjects of *kinematics* and *kinetics*. Kinematics is the study of the geometry and time evolution of motion independent of the forces causing the motion, while kinetics concerns itself with the interplay between forces and motion.

Statics as a subject has its origin in antiquity. Dynamics, in comparison, is a much younger discipline. The first systematic studies in dynamics were undertaken by Galileo Galilei (1564-1642). With the help of a series of brilliantly designed experiments, he was able to determine the laws of motion governing bodies in free fall and those of projectiles, as well as the law of inertia in 1638. To fully appreciate Galilei's achievements, one should note that differential and integral calculus were unknown in his time and instruments to precisely measure time were non-existent.

The scientific foundations for dynamics were laid down by Isaac Newton (1643-1727), who in 1687 formulated what we now know as *Newton's Laws of Motion*. Newton's Laws were able to accurately explain all experimental evidence at that time and the conclusions drawn from them have been confirmed to accurately predict the motion of all macroscopic bodies. We will treat these laws as axiomatic in character – they are not subject to mathematical proof.

Before we can study the interplay of forces and motion, it will be useful to first consider the purely geometric aspects of motion (kinematics). In this regard, we will carefully introduce the notions trajectory, velocity, and acceleration. Depending upon the type of motion (e.g. rectilinear, planar, or three-dimensional) we will describe these concepts using a variety of variables and coordinate systems. Our point of departure for the study of dynamics will be Newton's Laws of Motion. We will restrict our attention to the study of point masses and rigid bodies. With the help of these idealizations, we will see that we can effectively treat a wide variety of complex technical problems and arrive at useful solutions.

Newton's Laws of Motion are valid only in *inertial frames of reference*. However, it is often more convenient to formulate problems relative to *moving* frames of reference. In this regard, we will also briefly treat the topic of relative motion.

Newton's Laws of Motion are equivalent to the so-called *principles of mechanics* – the virtual power or work principles. In the solution of some problems it is useful to employ these alternate forms of the fundamental laws. We will restrict ourselves to the presentation of d'Alembert's principle and Lagrange equations of the 2nd type.

In the study of dynamics we will reuse many concepts we have already introduced in the study of statics, e.g. space, mass, force, moment, and idealizations such as point masses, rigid bodies, and point forces. Fundamental concepts from statics such as section cuts, the action-reaction law, and the force parallelogram law will also be employed. In the solution of concrete problems, we will also see that free-body diagrams will play a central role, just as they did in the study of statics. For the study of motion, we will further see that we will have to introduce a new fundamental variable, time, which was unnecessary in statics. With the introduction of time, we will find the need to define new dynamical concepts (e.g. velocity, acceleration, impulse, kinetic energy) and dynamical laws (e.g. impulse law and the work-energy theorem); it is with these concepts and related ideas that we will occupy ourselves in the chapters to follow.

Chapter 1

Motion of a Point Mass

1

1 Motion of a Point Mass

———— Objectives: We will first learn how one describes the motion of a point mass by its position, velocity, and acceleration in different coordinate systems and how such quantities can be determined. Subsequently, we will concern ourselves with the equations of motion, which prescribe the relation between forces and motion. An important role will again be played by the free-body diagram with whose help we will be able to properly derive the equations of motion. Further, we will discuss important physical concepts such as momentum, angular momentum, and work-laws and their applications.

1.1 Kinematics

1.1.1 Velocity and Acceleration

The subject of kinematics is the description of motion in space. Kinematics can be thought of as the geometry of motion independent of the cause of the motion.

The position of a point mass M in space is given by a point P and is uniquely described by its *position vector* r (Fig. 1.1a). This vector shows the momentary or instantaneous location of M relative to a fixed reference point in space, 0. If M changes location with time t, then $r(t)$ describes the *trajectory* or *path* of M.

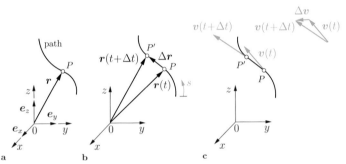

Fig. 1.1

Let us now consider two neighboring locations for M, P and P', at two different times t and $t + \Delta t$ (Fig. 1.1b). The change in the position vector over the time interval Δt is given by $\Delta r = r(t + \Delta t) - r(t)$. The *velocity* of M is defined as the limit of the change in position with respect to time:

$$v = \lim_{\Delta t \to 0} \frac{r(t + \Delta t) - r(t)}{\Delta t} = \lim_{\Delta t \to 0} \frac{\Delta r}{\Delta t} = \frac{dr}{dt} = \dot{r}. \quad (1.1)$$

Thus, the velocity v is the time derivative of the position vector r. We will usually denote time derivatives with a superposed dot.

Velocity is a vector. Since the change of the position vector, Δr, in the limit as $\Delta t \to 0$ points in the direction of the tangent to the trajectory of M, the velocity is always *tangent* to this curve. The velocity points in the direction that the mass traverses the path in space. In order to determine the magnitude of the velocity vector, we introduce the *arc-length s* as a measure of distance covered by M along its trajectory. Assume that the mass has moved a distance s up to the location P and a distance $s + \Delta s$ up to the location P'. With $|\Delta r| = \Delta s$, one obtains from (1.1) the *speed*

$$|v| = v = \lim_{\Delta t \to 0} \frac{\Delta s}{\Delta t} = \frac{ds}{dt} = \dot{s}. \tag{1.2}$$

Velocity and speed have dimensions of distance/time and are often measured in units of m/s. The units of km/h, which are used in transportation applications, are related as $1\,\text{km/h} = \frac{1000}{3600}\,\text{m/s} = \frac{1}{3.6}\,\text{m/s}$ or $1\,\text{m/s} = 3.6\,\text{km/h}$.

In general, velocity depends on time. In neighboring positions P and P' (Fig. 1.1c) the considered point mass has velocities $v(t)$ and $v(t + \Delta t)$. Thus, the change in the velocity is given by $\Delta v = v(t + \Delta t) - v(t)$. The *acceleration* is defined as the limit of this change with respect to time:

$$a = \lim_{\Delta t \to 0} \frac{v(t + \Delta t) - v(t)}{\Delta t} = \lim_{\Delta t \to 0} \frac{\Delta v}{\Delta t} = \frac{dv}{dt} = \dot{v} = \ddot{r}. \tag{1.3}$$

Thus the acceleration a is the first derivative of v and the second derivative of r. Acceleration is a vector. But since Δv (see Fig. 1.1c) does not have an obvious relation to the trajectory, we can not easily make statements about its direction and magnitude. Acceleration has dimensions of distance/time^2 and is often measured in units of m/s^2.

Velocity and acceleration have been introduced independent of a coordinate system. However, to solve specific problems, it is useful to introduce particular coordinates. In what follows, we will consider three important coordinate systems.

1.1.2 Velocity and Acceleration in Cartesian Coordinates

If we want to describe motion in Cartesian coordinates, we can choose 0 as the origin of a fixed (in space) system x, y, z. With unit vectors (basis vectors) $\boldsymbol{e}_x, \boldsymbol{e}_y, \boldsymbol{e}_z$ in the three coordinate directions (Fig. 1.1a), the position vector is given as

$$\boldsymbol{r}(t) = x(t)\,\boldsymbol{e}_x + y(t)\,\boldsymbol{e}_y + z(t)\,\boldsymbol{e}_z\,. \tag{1.4}$$

This is a parametric description of the trajectory with t as the parameter. If one can eliminate time from the three component relations in (1.4), then one has a time independent geometric description of the trajectory (cf. e.g. Section 1.2.2).

Using (1.1), one finds the velocity via differentiation (the basis vectors do not depend on time):

$$\boldsymbol{v} = \dot{\boldsymbol{r}} = \dot{x}\,\boldsymbol{e}_x + \dot{y}\,\boldsymbol{e}_y + \dot{z}\,\boldsymbol{e}_z\,. \tag{1.5}$$

Further differentiation gives the acceleration as

$$\boldsymbol{a} = \dot{\boldsymbol{v}} = \ddot{\boldsymbol{r}} = \ddot{x}\,\boldsymbol{e}_x + \ddot{y}\,\boldsymbol{e}_y + \ddot{z}\,\boldsymbol{e}_z\,. \tag{1.6}$$

Thus the components of the velocity and acceleration in Cartesian coordinates are given as

$$\begin{aligned} v_x &= \dot{x}, & v_y &= \dot{y}, & v_z &= \dot{z}, \\ a_x &= \dot{v}_x = \ddot{x}, & a_y &= \dot{v}_y = \ddot{y}, & a_z &= \dot{v}_z = \ddot{z}. \end{aligned} \tag{1.7}$$

The magnitudes follow as

$$v = \sqrt{\dot{x}^2 + \dot{y}^2 + \dot{z}^2} \quad \text{and} \quad a = \sqrt{\ddot{x}^2 + \ddot{y}^2 + \ddot{z}^2}. \tag{1.8}$$

1.1.3 Rectilinear Motion

Rectilinear motion is the simplest form of motion. Even so, it has many practical uses. For example, the free fall of a body in the earth's gravitational field or the travel of a train over a bridge are rectilinear motions.

If a point mass M moves along a straight line, then we can assume without loss of generality that the x-axis is coincident with this line (Fig. 1.2). Then according to (1.4), the position vector $\boldsymbol{r}$ to its current location P only has an x-component – likewise for the velocity $\boldsymbol{v}$ and the acceleration $\boldsymbol{a}$ according to (1.5) and (1.6). Thus, we can dispense with the vector character of the position, velocity, and acceleration and obtain from (1.7)

$$v = \dot{x}, \qquad a = \dot{v} = \ddot{x}. \tag{1.9}$$

In the case that v or a is negative, this means that the velocity respectively the acceleration is in the negative x-direction. An acceleration that decreases the magnitude of the velocity is known as a "deceleration".

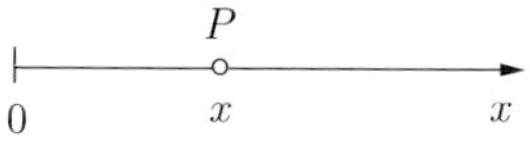

$$\begin{array}{cccc} & P & & \\ \vdash\!\!\!&\!\!\!\underset{x}{\circ}\!\!\!&\!\!\!\longrightarrow & \\ 0 & x & & x \end{array}$$

Fig. 1.2

In a case of rectilinear motion, if the position x is known as a function of time t, then the velocity and acceleration can be found via differentiation as indicated in (1.9). Often, however, problems are encountered where the acceleration is known and the velocity and position need to be determined. In these cases, integration is needed – a situation that is in general mathematically more difficult than differentiation. The determination of kinematic unknowns from given kinematic variables constitute basic kinematic problems. We take up these basic questions in what follows, where we will restrict ourselves to the most important special cases – those where the given kinematic variable depends on only one other variable. If the acceleration is taken as the given variable, there are five basic kinematic problems which we would like to treat.

1. $\boxed{a = 0}$ If the acceleration is zero, then according to (1.9) $a = \dot{v} = dv/dt = 0$. Integration then says that the velocity is constant:

$$v = \text{const} = v_0.$$

A motion with a constant velocity is known as a *uniform motion*. The position x can be found from $v = v_0 = \mathrm{d}x/\mathrm{d}t$ via integration. To do so, a statement about the start of the motion is needed, a so-called *initial condition*. Let us denote initial values by the subscript 0, so that at time $t = t_0$ the position is assumed as $x = x_0$. With integration we can follow two procedures:

a) *Indeterminate Integration.* After separation of variables, $\mathrm{d}x = v_0 \, \mathrm{d}t$, indeterminate integration gives

$$\int \mathrm{d}x = \int v_0 \, \mathrm{d}t \quad \rightarrow \quad x = v_0 \, t + C_1 \,.$$

The constant of integration C_1 is determined by exploiting the initial value:

$$x_0 = v_0 \, t_0 + C_1 \quad \rightarrow \quad C_1 = x_0 - v_0 \, t_0 \,.$$

Thus the desired position as a function of time is

$$x = x_0 + v_0 \, (t - t_0) \,.$$

b) *Determinate Integration.* After separation of variables, $\mathrm{d}x = v_0 \, \mathrm{d}t$, a determinate integration (where the lower integration limits are the initial values t_0, x_0) gives

$$\int_{x_0}^{x} \mathrm{d}\bar{x} = \int_{t_0}^{t} v_0 \, \mathrm{d}\bar{t} \quad \rightarrow \quad x - x_0 = v_0 \, (t - t_0)$$

or

$$x = x_0 + v_0 \, (t - t_0) \,.$$

Note that the variables under the integral sign are denoted with a bar, so that they are not confused with the upper limits of integration.

In what follows we will alternatively use both integration methods. Thus the initial conditions will either be used to determine constants of integration or will be used to set the lower limits of integration.

2. $\boxed{a = a_0}$ A motion with a constant acceleration is called a *uniform acceleration*. Let us assume that $t_0 = 0$ and the initial conditions for the velocity and position are

$$\dot{x}(0) = v_0, \qquad x(0) = x_0.$$

Then via integration of (1.9) the velocity and position follow as

$$\mathrm{d}v = a_0\,\mathrm{d}t \quad \rightarrow \quad \int_{v_0}^{v} \mathrm{d}\bar{v} = \int_{0}^{t} a_0\,\mathrm{d}\bar{t} \quad \rightarrow \quad v = v_0 + a_0\,t\,,$$

and

$$\mathrm{d}x = v\,\mathrm{d}t \quad \rightarrow \quad \int_{x_0}^{x} \mathrm{d}\bar{x} = \int_{0}^{t} (v_0 + a_0\,\bar{t})\,\mathrm{d}\bar{t}$$

$$\rightarrow \quad x = x_0 + v_0\,t + a_0\,\frac{t^2}{2}\,.$$

Fig. 1.3 shows the acceleration a, velocity v and position x as functions of time. One sees from the graphs, that a constant acceleration a_0 leads to a linear velocity $a_0 t + v_0$ and to a quadratic position-time dependency $a_0 t^2/2 + v_0 t + x_0$.

In nature, for example, one encounters uniform acceleration during *free fall* and other vertical motions in the earth's gravitational field. Galilei (1564–1642) discovered in 1638, that all bodies (ignoring air resistance) have a constant acceleration during free fall. This acceleration is called the *earth's gravitational acceleration* g. At the earth's surface, it has the value $g = 9.81$ m/s^2, where small variations with geographical latitude are neglected.

In what follows we will examine free fall and other vertical motions of a body K. As shown in Fig. 1.4a, we introduce a z-coordinate axis perpendicular to the earth's surface with the positive direction taken as upwards. For initial conditions, assume that

$$\dot{z}(0) = v_0, \qquad z(0) = z_0\,.$$

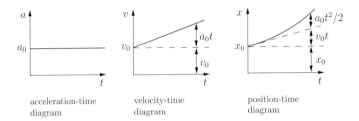

acceleration-time
diagram

velocity-time
diagram

position-time
diagram

Fig. 1.3

Then taking into account the sign of the earth's acceleration (in the *negative z*-direction), we obtain

$$\ddot{z} = a = -g,$$
$$\dot{z} = v = -g\,t + v_0, \qquad (1.10)$$
$$z = -\frac{g\,t^2}{2} + v_0\,t + z_0.$$

Let us now consider the special case of free fall. The body is dropped from a height $z_0 = H$ with zero initial velocity $(v_0 = 0)$. Then from (1.10)

$$a = -g, \qquad v = -g\,t, \qquad z = -\frac{g\,t^2}{2} + H.$$

If we wish to determine the time T that it takes for the body to fall to the ground, then we simply have to set $z = 0$ to find

$$z = 0 = -\frac{g\,T^2}{2} + H \quad \rightarrow \quad T = \sqrt{\frac{2\,H}{g}}.$$

If we insert this time into the expression for the velocity, then we will find the impact velocity

$$v_I = v(T) = -g\,T = -g\sqrt{\frac{2\,H}{g}} = -\sqrt{2\,gH}.$$

The minus sign shows that the velocity is oriented in the direction opposite to the positive z-direction (i.e., the negative z-direction).

Fig. 1.4b shows the position of the body as a function of time.

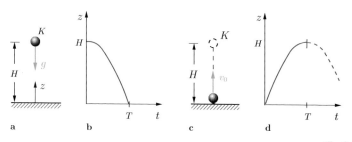

Fig. 1.4

We will now examine the vertical motion of a body that is thrown vertically at time $t = 0$ from the earth's surface $(z_0 = 0)$ with initial velocity v_0 (Fig. 1.4c). It follows from (1.10), that

$$a = -g, \quad v = -gt + v_0, \quad z = -\frac{gt^2}{2} + v_0 t.$$

The body will reach its highest point (H), when the velocity is zero. The time that it takes to reach this point follows as

$$v = 0 = -gT + v_0 \quad \rightarrow \quad T = \frac{v_0}{g}.$$

Substituting this time into the expression for position, one finds the value for the highest point on the trajectory:

$$H = z(T) = -\frac{gT^2}{2} + v_0 T = -\frac{g}{2}\frac{v_0^2}{g^2} + v_0\frac{v_0}{g} = \frac{v_0^2}{2g}.$$

Fig. 1.4d shows the dependency of the body's position with time. Comparing this result with that of free fall, one sees the close relationship between the two motions: a body that falls from a height H hits the ground with a velocity $|v_I| = \sqrt{2gH}$, whereas a body that is thrown vertically with a velocity v_0 reaches a height of $H = v_0^2/2g$.

3. $a = a(t)$ In this case, the velocity v and the position x can be directly found via two successive integrations of (1.9) with respect

to time. With initial conditions $v(t_0) = v_0$, $x(t_0) = x_0$ one has

$$\mathrm{d}v = a(t)\,\mathrm{d}t \quad \rightarrow \quad v = v_0 + \int_{t_0}^{t} a(\bar{t})\,\mathrm{d}\bar{t}, \tag{1.11}$$

$$\mathrm{d}x = v(t)\,\mathrm{d}t \quad \rightarrow \quad x = x_0 + \int_{t_0}^{t} v(\bar{t})\,\mathrm{d}\bar{t}. \tag{1.12}$$

4. **$a = a(v)$** In this case the acceleration is a given function of the velocity. Thus from (1.9) via *separation of variables*

$$a(v) = \frac{\mathrm{d}v}{\mathrm{d}t} \quad \rightarrow \quad \mathrm{d}t = \frac{\mathrm{d}v}{a(v)}.$$

Determinate integration (with the lower limit as the initial condition $t = t_0$, $v(t_0) = v_0$) gives

$$\int_{t_0}^{t} \mathrm{d}\bar{t} = \int_{v_0}^{v} \frac{\mathrm{d}\bar{v}}{a(\bar{v})} \quad \rightarrow \quad t = t_0 + \int_{v_0}^{v} \frac{\mathrm{d}\bar{v}}{a(\bar{v})} = f(v). \tag{1.13}$$

In this manner, the time t is given as a function of the velocity v. If one can solve this relation to find $v = F(t)$ (i.e., find the inverse function $F = f^{-1}$), then the position can be determined from (1.12) as

$$x = x_0 + \int_{t_0}^{t} F(\bar{t})\,\mathrm{d}\bar{t}. \tag{1.14}$$

In this way, the position x is now given as a function of time t.

From $a(v)$ one can directly determine the position x as a function of v. Using the chain rule

$$a = \frac{\mathrm{d}v}{\mathrm{d}t} = \frac{\mathrm{d}v}{\mathrm{d}x}\frac{\mathrm{d}x}{\mathrm{d}t} = \frac{\mathrm{d}v}{\mathrm{d}x} v$$

and applying separation of variables gives

$$\mathrm{d}x = \frac{v}{a}\,\mathrm{d}v.$$

Determinate integration using the initial values v_0 and x_0 gives

$$x = x_0 + \int\limits_{v_0}^{v} \frac{\bar{v}}{a(\bar{v})} \, d\bar{v}. \tag{1.15}$$

As an illustrative example, let us consider the motion of a point mass whose acceleration $a = -kv$, where k is a given constant. Such motions occur, for example, for bodies moving in viscous fluids (cf. Section 1.2.4). Let $x(0) = x_0$ and $v(0) = v_0$ be given as initial conditions.

From (1.13) it follows that

$$t = \int\limits_{v_0}^{v} \frac{d\bar{v}}{-k\bar{v}} = -\frac{1}{k} \ln \bar{v} \Big|_{v_0}^{v} = -\frac{1}{k} \ln \frac{v}{v_0} = f(v).$$

Solving for v (determining the inverse function), gives

$$v = v_0 \, e^{-kt} = F(t).$$

Then from (1.14), one finds

$$x = x(t) = x_0 + \int\limits_{0}^{t} v_0 \, e^{-k\bar{t}} \, d\bar{t} = x_0 + \left(-\frac{v_0}{k} \right) e^{-k\bar{t}} \Big|_{0}^{t}$$

$$= x_0 + \frac{v_0}{k} (1 - e^{-kt}).$$

Using instead (1.15), we alternately have

$$x = x(v) = x_0 + \int\limits_{v_0}^{v} \frac{\bar{v}}{-k\bar{v}} \, d\bar{v} = x_0 - \frac{1}{k} (v - v_0).$$

If we substitute the velocity $v = v_0 \, e^{-kt}$, then we recover the previously determined position-time relation

$$x = x_0 - \frac{1}{k} (v_0 \, e^{-kt} - v_0) = x_0 + \frac{v_0}{k} (1 - e^{-kt}) = x(t).$$

The result is shown in Fig. 1.5. Since the acceleration a is proportional to $-v$, the point continuously decelerates. Thus the

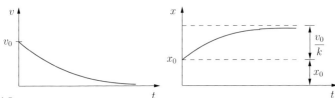

Fig. 1.5

velocity v continuously decreases and with smaller magnitude v the deceleration also decreases. Only in the limit $t \to \infty$ does the velocity reach zero. The position of the point asymptotically approaches $x_0 + v_0/k$. This value follows from $x(t)$ in the limit as $t \to \infty$ or from $x(v)$ as $v \to 0$.

5. $\boxed{a = a(x)}$ Let us once again use the chain rule

$$a = \frac{\mathrm{d}v}{\mathrm{d}t} = \frac{\mathrm{d}v}{\mathrm{d}x}\frac{\mathrm{d}x}{\mathrm{d}t} = \frac{\mathrm{d}v}{\mathrm{d}x}v$$

and separation of variables:

$$v\,\mathrm{d}v = a\,\mathrm{d}x. \tag{1.16}$$

Integration with initial conditions $v(t_0) = v_0$, $x(t_0) = x_0$ gives

$$\frac{1}{2}v^2 = \frac{1}{2}v_0^2 + \int_{x_0}^{x} a(\bar{x})\,\mathrm{d}\bar{x} = f(x) \quad \rightarrow \quad v = \sqrt{2\,f(x)}. \tag{1.17}$$

In this way, we determine the dependency of velocity v in terms of position x. From the relation $v = \mathrm{d}x/\mathrm{d}t$, one finds after applying separation of variables and integrating, that

$$\mathrm{d}t = \frac{\mathrm{d}x}{v} = \frac{\mathrm{d}x}{\sqrt{2\,f(x)}} \quad \rightarrow \quad t = t_0 + \int_{x_0}^{x} \frac{\mathrm{d}\bar{x}}{\sqrt{2\,f(\bar{x})}} = g(x). \tag{1.18}$$

Thus, time t is now known as a function of position. If one can invert the relation $t = g(x)$ to yield $x = G(t)$, then one also obtains position as a function of time.

As an example, let us study a motion with the given acceleration relation $a = -\omega^2 x$, where ω^2 is a given constant. At time $t_0 = 0$ assume $x(0) = x_0$ and $v(0) = v_0 = 0$. Substitution into (1.17) gives

$$\frac{1}{2}v^2 = \int_{x_0}^{x}(-\omega^2\,\bar{x})\,\mathrm{d}\bar{x} = -\omega^2\left(\frac{x^2}{2} - \frac{x_0^2}{2}\right) = \frac{\omega^2}{2}(x_0^2 - x^2) = f(x)$$

$$\rightarrow \quad v = \pm\omega\sqrt{x_0^2 - x^2}\,.$$

From (1.18) one obtains the time response as

$$t = t(x) = \pm\int_{x_0}^{x}\frac{\mathrm{d}\bar{x}}{\omega\sqrt{x_0^2 - \bar{x}^2}} = \pm\frac{1}{\omega}\arcsin\frac{\bar{x}}{x_0}\Big|_{x_0}^{x}$$

$$= \pm\frac{1}{\omega}\left(\arcsin\frac{x}{x_0} - \frac{\pi}{2}\right) = \pm\frac{1}{\omega}\arccos\frac{x}{x_0}\,.$$

Inverting, one then finds the position as a function of time:

$$x = x_0\cos\omega t.$$

This motion is a *harmonic oscillation* (cf. Chapter 5). By differentiation, one can also obtain the velocity and acceleration as functions of time:

$$v(t) = \dot{x} = -\omega x_0\sin\omega t, \quad a(t) = \ddot{x} = -\omega^2 x_0\cos\omega t = -\omega^2 x(t)\,.$$

Fig. 1.6a displays the mass's position and velocity as functions of time.

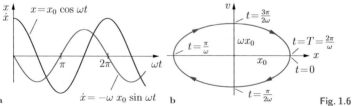

Fig. 1.6

Often one is also interested in the dependency of the velocity on the position. We can geometrically display this dependency in an x, v-diagram as a *phase curve* or *phase trajectory*.

For the oscillation example, we have $v = \pm\omega\sqrt{x_0^2 - x^2}$. Thus it follows, that

$$v^2 = \omega^2(x_0^2 - x^2) \quad \rightarrow \quad \left(\frac{x}{x_0}\right)^2 + \left(\frac{v}{\omega x_0}\right)^2 = 1\,.$$

The phase curve is an ellipse with semi-axes x_0 and ωx_0 (Fig. 1.6b). Each point x, v on the phase curve corresponds to particular time points t: time is a parameter. Since the curve is closed in this example, the motion repeats itself after each pass (oscillation = periodic event). The figure shows distinct times and the direction of the motion. The time it takes to complete a single cycle, $T = 2\pi/\omega$, is known as the period of oscillation or for short the period (cf. Chapter 5).

In other cases when velocity and position are known functions of time, one needs to eliminate time from the relations $\dot{x}(t)$ and $x(t)$ in order to determine the phase curve.

As closure to the developments up to this point, Table 1.1 summarizes the important relations associated with the basic kinematic questions.

Table 1.1

Given	Sought	
$a(t)$	$v = v_0 + \int_{t_0}^{t} a(\bar{t})\,\mathrm{d}\bar{t}$	$x = x_0 + \int_{t_0}^{t} v(\bar{t})\,\mathrm{d}\bar{t}$
$a(v)$	$t = t_0 + \int_{v_0}^{v} \dfrac{\mathrm{d}\bar{v}}{a(\bar{v})}$	$x = x_0 + \int_{v_0}^{v} \dfrac{\bar{v}\,\mathrm{d}\bar{v}}{a(\bar{v})}$
$a(x)$	$v^2 = v_0^2 + 2\int_{x_0}^{x} a(\bar{x})\,\mathrm{d}\bar{x}$	$t = t_0 + \int_{x_0}^{x} \dfrac{\mathrm{d}\hat{x}}{\sqrt{v_0^2 + 2\int_{x_0}^{\hat{x}} a(\bar{x})\,\mathrm{d}\bar{x}}}$

E1.1

Example 1.1 A vehicle travelling on a straight path has, at time $t_0 = 0$, a velocity $v_0 = 40$ m/s and acceleration $a_0 = 5$ m/s^2. It then experiences a linearly decreasing acceleration in time to a value of $a = 0$ at $t = 6$ s. Next, it travels a distance $s_2 = 550$ m in uniform motion and finally in a third phase of travel it is uniformly decelerated with an acceleration $|a_3| = 11$ m/s^2 until it stops.

At what time and in what location does the vehicle come to a stop? Sketch the acceleration-, velocity-, and position-time diagrams.

Solution For each of the three time intervals of the motion, we will introduce a new time variable (Fig. 1.7a). Variables at the end of a time interval will be denoted with a star. The position x will be computed relative to the location of the vehicle at time $t_0 = 0$.

1. Linear acceleration $(0 \leqq t_1 \leqq t_1^*)$.
The time variation of the acceleration can be represented as $a_1 = a_0(1-t_1/t_1^*)$. Considering the initial conditions $x_1(t_1=0) = 0$ and $v_1(t_1 = 0) = v_0$ one obtains from (1.11) and (1.12) the velocity

$$v_1 = v_0 + \int_0^{t_1} a_0 \left(1 - \frac{\bar{t}_1}{t_1^*} \right) d\bar{t}_1 = v_0 + a_0 \left(t_1 - \frac{t_1^2}{2\,t_1^*} \right)$$

and position

$$x_1 = \int_0^{t_1} v_1 \, d\bar{t}_1 = v_0\,t_1 + a_0 \left(\frac{t_1^2}{2} - \frac{t_1^3}{6\,t_1^*} \right).$$

At the end of this interval of the motion $(t_1 = t_1^* = 6\,\text{s})$ we have

$$v_1^* = v_0 + a_0 \frac{t_1^*}{2} = 40 + 5 \cdot 3 = 55\,\frac{\text{m}}{\text{s}},$$

$$x_1^* = v_0\,t_1^* + a_0 \frac{t_1^{*2}}{3} = 40 \cdot 6 + 5 \cdot \frac{6^2}{3} = 300\,\text{m}.$$

2. Uniform motion $(0 \leqq t_2 \leqq t_2^*)$.
In the second interval of the motion, the velocity is a constant with value $v_2 = v_1^* = 55$ m/s. Thus the position is given by

$$x_2 = x_1^* + v_2\, t_2\,.$$

At time t_2^* the vehicle has traveled a total distance of

$$x_2^* = x_1^* + s_2 = 300 + 550 = 850\,\text{m}\,.$$

From $s_2 = v_2\, t_2^* = 550$ m, follows the time

$$t_2^* = \frac{s_2}{v_2} = \frac{550}{55} = 10\,\text{s}\,.$$

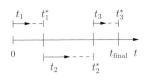

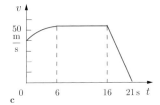

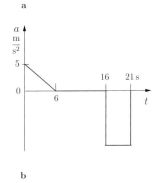

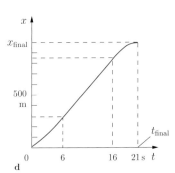

Fig. 1.7

3. Constant deceleration $(0 \leqq t_3 \leqq t_3^*)$.
The final conditions of the second time interval $(x_2^*, v_2^* = v_2)$ are the initial conditions for the third time interval. Thus we find $(a_3$

is the magnitude of the deceleration)

$$v_3 = v_2^* - a_3\, t_3,$$

$$x_3 = x_2^* + v_2^*\, t_3 - a_3\, \frac{t_3^2}{2}.$$

The time at which the vehicle comes to a stop follows from

$$v_3^* = v_2^* - a_3\, t_3^* = 0 \quad \rightarrow \quad t_3^* = \frac{v_2^*}{a_3} = \frac{55}{11} = 5\,\mathrm{s}\,,$$

and the final position is obtained as

$$\underline{x_{\mathrm{final}} = x_3^*} = x_2^* + v_2^*\, t_3^* - a_3\, \frac{t_3^{*2}}{2}$$

$$= 850 + 55 \cdot 5 - 11 \cdot \frac{5^2}{2} = \underline{987.5\,\mathrm{m}}\,.$$

The total time of travel is

$$\underline{t_{\mathrm{final}}} = t_1^* + t_2^* + t_3^* = 6 + 10 + 5 = \underline{21\,\mathrm{s}}\,.$$

Figs. 1.7b-d show the acceleration-, velocity-, and position-time diagrams. At the time where the acceleration has a jump, the velocity has a change in slope. In the position-time curve there is no such change in slope as the vehicle has not experienced any jumps in velocity. (Jumps in v only occur during impacts, cf. Section 2.5).

Example 1.2 A point mass M with current position P moves according to Fig. 1.8a along a straight line. The square of the velocity decreases linearly with x. The mass passes the location $x = 0$ at $t = 0$ with a velocity $v_0 > 0$ and at the location $x = x_1 > 0$ it has the velocity $v_1 = 0$.

At what time does the point mass reach the position x_1 and what is its acceleration?

Solution First, we need to describe the velocity. A linear relation between v^2 and x can be generally written as $v^2 = b\,x + c$. The

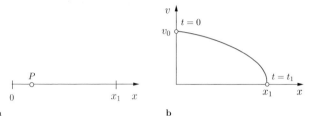

Fig. 1.8   a   b

constants b and c follow from consideration of the given values:

$$v(x=0) = v_0 \;\rightarrow\; c = v_0^2, \quad v(x=x_1) = 0 \;\rightarrow\; b = -\frac{c}{x_1} = -\frac{v_0^2}{x_1}.$$

Thus,

$$v = v(x) = v_0\sqrt{1 - \frac{x}{x_1}}\,.$$

Figure 1.8b shows $v(x)$ in the phase plane. From $v = \mathrm{d}x/\mathrm{d}t$ it follows by separation of variables and indeterminate integration that

$$t = \int \frac{\mathrm{d}x}{v_0\sqrt{1 - \frac{x}{x_1}}} = -2\,\frac{x_1}{v_0}\sqrt{1 - \frac{x}{x_1}} + C.$$

The integration constant C is determined from the initial condition $x(0) = 0$:

$$0 = -2\,\frac{x_1}{v_0}\sqrt{1} + C \quad\rightarrow\quad C = 2\,\frac{x_1}{v_0}\,.$$

With this, the time t_1 when the mass reaches $x = x_1$ is found as

$$\underline{\underline{t_1}} = t(x_1) = C = \underline{\underline{2\,\frac{x_1}{v_0}}}\,.$$

The acceleration is determined by application of the chain rule:

$$\underline{\underline{a}} = \frac{\mathrm{d}v}{\mathrm{d}t} = \frac{\mathrm{d}v}{\mathrm{d}x}\frac{\mathrm{d}x}{\mathrm{d}t} = \frac{\mathrm{d}v}{\mathrm{d}x}\,v = -\frac{v_0}{2\,x_1}\frac{1}{\sqrt{1 - \frac{x}{x_1}}}\,v_0\sqrt{1 - \frac{x}{x_1}} = \underline{\underline{-\frac{v_0^2}{2\,x_1}}}\,.$$

Since $a = \text{const}$, we see that the motion is a uniform acceleration.

As a check, we can compute the velocity and position via integration from the known acceleration:

$$v = a\,t + v_0 = -\frac{v_0^2}{2\,x_1}\,t + v_0,$$

$$x = -\frac{v_0^2}{4\,x_1}\,t^2 + v_0\,t.$$

Elimination of t leads to the given velocity-position relation.

E1.3

Example 1.3 A point mass M with current position P moves, as shown in Fig. 1.9, along the x-axis with acceleration $a = k\,\sqrt{v}$, where the constant $k = 2\,(\mathrm{m/s^3})^{1/2}$. At time $t = 0$, M passes the location $x_0 = 1/3\,\mathrm{m}$ with a velocity $v_0 = 1\,\mathrm{m/s}$.

Find the location x_1 of M at time $t_1 = 2\,\mathrm{s}$. What are the velocity and acceleration at this time?

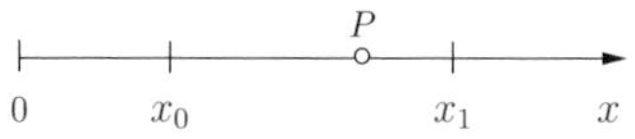

Fig. 1.9

Solution The acceleration is given as a function of velocity. According to (1.13), we have

$$t = \int_{v_0}^{v} \frac{\mathrm{d}\bar{v}}{k\sqrt{\bar{v}}} = \frac{2}{k}(\sqrt{v} - \sqrt{v_0}) \quad \rightarrow \quad v = v(t) = \left(\frac{k\,t}{2} + \sqrt{v_0}\right)^2.$$

Indeterminate integration of v gives

$$x = \int v\,\mathrm{d}t = \frac{1}{3}\frac{2}{k}\left(\frac{k\,t}{2} + \sqrt{v_0}\right)^3 + C.$$

The integration constant C is found using the initial condition $x(0) = x_0$:

$$\frac{1}{3} = \frac{1}{3}\frac{2}{2}(\sqrt{1})^3 + C \quad \rightarrow \quad C = 0.$$

Thus, we arrive at the result

$$x = \frac{2}{3\,k}\left(\frac{k\,t}{2} + \sqrt{v_0}\right)^3,$$

$$v = \dot{x} = \left(\frac{k\,t}{2} + \sqrt{v_0}\right)^2,$$

$$a = \dot{v} = \ddot{x} = k\left(\frac{k\,t}{2} + \sqrt{v_0}\right).$$

At time $t = t_1$ we find

$$\underline{\underline{x_1}} = x\,(t_1) = \frac{2}{3\cdot 2}\left(\frac{2\cdot 2}{2} + \sqrt{1}\right)^3 = \underline{\underline{9\,\mathrm{m}}},$$

$$\underline{\underline{v_1}} = v\,(t_1) = \left(\frac{2\cdot 2}{2} + \sqrt{1}\right)^2 = \underline{\underline{9\,\frac{\mathrm{m}}{\mathrm{s}}}},$$

$$\underline{\underline{a_1}} = a\,(t_1) = 2\left(\frac{2\cdot 2}{2} + \sqrt{1}\right) = \underline{\underline{6\,\frac{\mathrm{m}}{\mathrm{s}^2}}}.$$

As a check, we can easily see that the result is compatible with the initially given acceleration formula: $a = k\,\sqrt{v}$.

1.1.4 Planar Motion, Polar Coordinates

When a point mass M moves in a plane (e.g. in the x, y-plane), one often finds it easiest to describe the motion using Cartesian coordinates according to relations (1.4) to (1.8) (ignoring the component orthogonal to the plane of motion). However, it is often also useful to describe its current position P using polar coordinates r, φ; see Fig. 1.10a. Let us introduce orthogonal basis vectors e_r and e_φ, such that e_r always points from the *fixed* point 0 to M. In this case, the position vector is

$$r = r\,e_r. \tag{1.19}$$

To find expressions for the velocity and acceleration we need to differentiate the position vector. Since the location of M changes with time, the directions of e_r and e_φ also change with

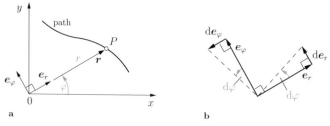

Fig. 1.10

time. In contradistinction to the fixed-in-space basis vectors in a Cartesian coordinate system, in polar coordinates the basis vectors must also be differentiated. The basis vector e_r is assumed to be a unit vector. Its change due to an infinitesimal rotation $d\varphi$ over a time dt gives according to Fig. 1.10b a vector de_r, which is orthogonal to e_r (i.e. it points in the direction of e_φ) and has a magnitude $1 \cdot d\varphi$. Thus we can write

$$de_r = d\varphi\, e_\varphi \quad \rightarrow \quad \dot{e}_r = \frac{de_r}{dt} = \frac{d\varphi}{dt} e_\varphi = \dot{\varphi}\, e_\varphi\,.$$

In a similar fashion, one can determine the change in the basis vector e_φ from Fig. 1.10b:

$$de_\varphi = -d\varphi\, e_r \quad \rightarrow \quad \dot{e}_\varphi = \frac{de_\varphi}{dt} = -\frac{d\varphi}{dt} e_r = -\dot{\varphi}\, e_r\,.$$

It then follows from (1.19) that the velocity is given by

$$\boldsymbol{v} = \dot{\boldsymbol{r}} = \dot{r}\, \boldsymbol{e}_r + r\, \dot{\boldsymbol{e}}_r = \dot{r}\, \boldsymbol{e}_r + r\dot{\varphi}\, \boldsymbol{e}_\varphi\,. \tag{1.20}$$

It has a *radial* component $v_r = \dot{r}$ and an *angular* component $v_\varphi = r\,\dot{\varphi}$. As noted before, the velocity is tangential to the trajectory.

Differentiation of (1.20) provides an expression for the acceleration:

$$\begin{aligned}\boldsymbol{a} = \dot{\boldsymbol{v}} &= \ddot{r}\, \boldsymbol{e}_r + \dot{r}\, \dot{\boldsymbol{e}}_r + \dot{r}\dot{\varphi}\, \boldsymbol{e}_\varphi + r\ddot{\varphi}\, \boldsymbol{e}_\varphi + r\dot{\varphi}\, \dot{\boldsymbol{e}}_\varphi \\ &= \left(\ddot{r} - r\dot{\varphi}^2\right) \boldsymbol{e}_r + \left(r\ddot{\varphi} + 2\,\dot{r}\dot{\varphi}\right) \boldsymbol{e}_\varphi\,. \end{aligned} \tag{1.21}$$

It has a radial component $a_r = \ddot{r} - r\dot{\varphi}^2$ and an angular component $a_\varphi = r\ddot{\varphi} + 2\,\dot{r}\dot{\varphi}$. As noted earlier, one can not easily make general statements about the orientation of the acceleration with respect to the trajectory.

In summary, we have the following relations for planar motion in polar coordinates:

$$
\begin{aligned}
\boldsymbol{r} &= r\,\boldsymbol{e}_r, \\[4pt]
\boldsymbol{v} &= v_r\,\boldsymbol{e}_r + v_\varphi\,\boldsymbol{e}_\varphi = \dot{r}\,\boldsymbol{e}_r + r\dot{\varphi}\,\boldsymbol{e}_\varphi, \\[4pt]
\boldsymbol{a} &= a_r\,\boldsymbol{e}_r + a_\varphi\,\boldsymbol{e}_\varphi = (\ddot{r} - r\dot{\varphi}^2)\,\boldsymbol{e}_r + (r\ddot{\varphi} + 2\,\dot{r}\dot{\varphi})\,\boldsymbol{e}_\varphi\,.
\end{aligned}
\tag{1.22}
$$

Over a time interval $\mathrm{d}t$ the position vector changes by an angle $\mathrm{d}\varphi$. The time rate of change of this angle, $\dot{\varphi} = \mathrm{d}\varphi/\mathrm{d}t$, is known as the *angular velocity*. It is usually denoted by the letter ω:

$$
\omega = \dot{\varphi}\,.
\tag{1.23}
$$

The angular velocity has dimensions of 1/time.

Differentiation of ω leads to the *angular acceleration*

$$
\dot{\omega} = \ddot{\varphi}\,.
\tag{1.24}
$$

The angular acceleration is denoted by many authors by the letter α but we will usually not adopt this convention. It has dimensions of $1/\text{time}^2$.

An important special case of planar motion is *circular motion*; see (Fig. 1.11a) where $r = \text{const}$. In this case

$$
\boldsymbol{r} = r\,\boldsymbol{e}_r, \quad \boldsymbol{v} = r\omega\,\boldsymbol{e}_\varphi, \quad \boldsymbol{a} = -\,r\omega^2\,\boldsymbol{e}_r + r\dot{\omega}\,\boldsymbol{e}_\varphi\,.
\tag{1.25}
$$

The velocity has only an angular component

$$
v = v_\varphi = r\omega,
\tag{1.26}
$$

which points in the direction tangent to the circular path of the point mass (Fig. 1.11b). The acceleration has a component in the tangential direction

$$
a_\varphi = r\dot{\omega}
\tag{1.27}
$$

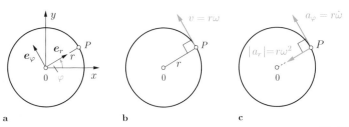

Fig. 1.11

and a component in the radial direction (orthogonal to the trajectory, Fig. 1.11c)

$$a_r = -r\omega^2 \,. \tag{1.28}$$

The minus sign indicates that the radial component points inwards – towards the origin. This component is called the *centripetal acceleration*.

If in addition to $r = $ const, the angular velocity $\omega = $ const, then the velocity has a constant magnitude $r\omega$, and the tangential acceleration is zero. In spite of this, there is still a radial acceleration of magnitude $r\omega^2$. It is needed to change the direction of the velocity vector.

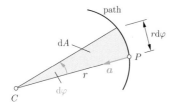

Fig. 1.12

Another important case of planar motion is *central motion*. In this case, the acceleration vector is assumed to continuously point towards *a single* point, the center C (Fig. 1.12). This occurs, for example, with the motion of the planets, where the sun serves as the center C. If we place the origin of our coordinate system at the center, then the angular component of the acceleration disappears:

$$a_\varphi = 0 \;\rightarrow\; r\dot{\omega} + 2\,\dot{r}\omega = \frac{1}{r}\frac{\mathrm{d}}{\mathrm{d}t}(r^2\,\omega) = 0 \;\rightarrow\; r^2\,\omega = \text{const}\,. \tag{1.29a}$$

We can give this result a clever interpretation. Fig. 1.12 shows that the ray r sweeps out an area $dA = \frac{1}{2}rrd\varphi$ in a time interval dt. Calculating the differential ratio

$$\frac{dA}{dt} = \frac{1}{2}r^2\frac{d\varphi}{dt} = \frac{1}{2}r^2\omega \tag{1.29b}$$

shows that the rate of change of this area in a central motion is constant (cf. (1.29a)). This observation is known as Kepler's 2nd Law of planetary motion (Friedrich Johannes Kepler, 1571-1630): the ray from the sun to a planet sweeps out equal areas in equal times. The result is also known as the *Law of Equal Areas*.

Example 1.4 A ship S moves as shown in Fig. 1.13a with a velocity of constant magnitude v, where the angle α between the velocity vector and the connecting line to the lighthouse L remains constant.

E1.4

   What is the magnitude of the acceleration and what is the trajectory of the ship?

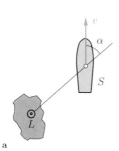
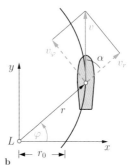

Fig. 1.13   a   b

Solution To describe the motion, we introduce polar coordinates with the origin at L (Fig. 1.13b). The velocity has in this case constant components

$$v_r = v\cos\alpha, \quad v_\varphi = v\sin\alpha\,.$$

Since $v_r = \dot{r}$ and $v_\varphi = r\dot{\varphi}$ (cf. (1.22)), it follows that

$$\dot{r} = v\cos\alpha, \quad \dot{\varphi} = \frac{v\sin\alpha}{r}\,.$$

Upon further differentiation, one finds

$$\ddot{r} = 0, \quad \ddot{\varphi} = \frac{d\dot{\varphi}}{dr}\dot{r} = -\frac{v\sin\alpha}{r^2}v\cos\alpha = -\frac{v^2\sin\alpha\cos\alpha}{r^2}.$$

Thus we find for the acceleration components

$$a_r = \ddot{r} - r\dot{\varphi}^2 = -\frac{v^2\sin^2\alpha}{r},$$

$$a_\varphi = r\ddot{\varphi} + 2\dot{r}\dot{\varphi} = -\frac{v^2\sin\alpha\cos\alpha}{r} + 2v\cos\alpha\frac{v\sin\alpha}{r}$$

$$= \frac{v^2\sin\alpha\cos\alpha}{r}.$$

The magnitude of the acceleration is then found as

$$\underline{\underline{a}} = \sqrt{a_r^2 + a_\varphi^2} = \frac{v^2\sin\alpha}{r}\sqrt{\sin^2\alpha + \cos^2\alpha} = \underline{\underline{\frac{v^2\sin\alpha}{r}}}.$$

The desired trajectory follows from

$$\dot{r} = v\cos\alpha \quad \rightarrow \quad dr = v\cos\alpha\,dt,$$

$$r\dot{\varphi} = v\sin\alpha \quad \rightarrow \quad rd\varphi = v\sin\alpha\,dt$$

via elimination of dt and separation of variables:

$$\frac{dr}{r} = \frac{d\varphi}{\tan\alpha}.$$

Indeterminate integration gives

$$\ln r = \frac{\varphi}{\tan\alpha} + C.$$

Assuming that the ship is a distance r_0 from the origin at angle $\varphi = 0$, then $C = \ln r_0$. Substituting, one finds

$$\ln r = \frac{\varphi}{\tan\alpha} + \ln r_0 \quad \rightarrow \quad \ln\frac{r}{r_0} = \frac{\varphi}{\tan\alpha}$$

or

$$\underline{\underline{r = r_0\,e^{\frac{\varphi}{\tan\alpha}}}}.$$

This is the expression for a logarithmic spiral.

Example 1.5 A flywheel (diameter $d = 60$ cm) is uniformly accelerated from a standstill such that at time $t_1 = 20$ s it has reached
a rotation rate of $n = 1000$ rpm (rpm = revolutions per minute).
 a) What is the magnitude of the angular acceleration $\dot{\omega}$ of the
flywheel? b) How many revolutions N has the flywheel made at
time t_1? c) What are the velocity and acceleration of a point on
the perimeter at time $t_2 = 1$ s?

Solution a)   For a uniformly accelerated circular motion the an-
gular acceleration is constant: $\dot{\omega} = \dot{\omega}_0 = \text{const}$. Noting the initial condition $\omega(0) = 0$, the angular velocity is $\omega = \dot{\omega}_0 t$. With
$\omega(t_1) = \omega_1$ it follows that

$$\dot{\omega}_0 = \frac{\omega_1}{t_1} \, .$$

Rotation rate n (in rpm) and angular velocity ω are easily related
to one another: with a rotation rate n one has an angle of revolution of $n \cdot 2\pi$ after one minute. Angular velocity is usually given
in units of $1/s$ (or equivalently rad/s); in this case

$$\omega = \frac{n \cdot 2\pi}{60} \, .$$

With the given rotation rate, it follows that

$$\underline{\underline{\dot{\omega}_0}} = \frac{1000 \cdot 2\pi}{60 \cdot 20} = \underline{\underline{5.24\,\text{s}^{-2}}} \, .$$

b) Integration of $\omega = \dot{\omega}_0 t$ with initial condition $\varphi(0) = 0$ gives
the rotation angle

$$\varphi = \frac{1}{2} \dot{\omega}_0 t^2 \, .$$

With the given values, we have at $t = t_1$ a rotation angle

$$\varphi_1 = \varphi(t_1) = \frac{1}{2} \cdot 5.24 \cdot 400 = 1048\,\text{rad} \, .$$

Thus the number of revolutions is

$$\underline{\underline{N}} = \frac{\varphi_1}{2\pi} = \underline{\underline{166}} \, .$$

c) From (1.26) the angular velocity component is obtained as

$$v = r\omega = r\dot{\omega}_0\, t\,.$$

The radial velocity is zero. From (1.27) and (1.28) the two acceleration components are

$$a_\varphi = r\dot{\omega}_0, \quad a_r = -r\omega^2 = -r(\dot{\omega}_0\, t)^2\,.$$

For the given numerical values, at $t = t_2$ we have

$$\underline{\underline{v}} = r\dot{\omega}_0\, t_2 = 30 \cdot 5.24 \cdot 1 = \underline{157.2\,\mathrm{cm/s}},$$

$$\underline{\underline{a_\varphi}} = 30 \cdot 5.24 = \underline{157.2\,\mathrm{cm/s}^2},$$

$$\underline{\underline{a_r}} = -30 \cdot (5.24 \cdot 1)^2 = \underline{-823.7\,\mathrm{cm/s}^2}$$

and

$$\underline{\underline{a}} = \sqrt{a_\varphi^2 + a_r^2} = \underline{838.6\,\mathrm{cm/s}^2}\,.$$

The centripetal acceleration a_r, which points towards the center of the flywheel, grows quadratically with t and is thus, after a short time, much larger than the time-independent angular acceleration a_φ.

1.1.5 Three-Dimensional Motion, Serret-Frenet Frame

The general motion of a point mass in three dimensions can be described with the previously introduced formulae either by Cartesian coordinates x, y, z or through cylindrical coordinates r, φ, z. By cylindrical coordinates, we mean a three-dimensional generalization of polar coordinates (Fig. 1.14), whereby the basis vector e_z is a constant, so that with (1.22) one has in cylindrical coordinates:

$$\boldsymbol{r} = r\,\boldsymbol{e}_r + z\,\boldsymbol{e}_z,$$

$$\boldsymbol{v} = \dot{r}\,\boldsymbol{e}_r + r\dot{\varphi}\,\boldsymbol{e}_\varphi + \dot{z}\,\boldsymbol{e}_z, \tag{1.30}$$

$$\boldsymbol{a} = (\ddot{r} - r\dot{\varphi}^2)\,\boldsymbol{e}_r + (r\ddot{\varphi} + 2\,\dot{r}\dot{\varphi})\,\boldsymbol{e}_\varphi + \ddot{z}\,\boldsymbol{e}_z\,.$$

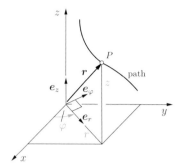

Fig. 1.14

Note that r is not the magnitude of the vector $\boldsymbol{r}$, but rather the magnitude of its projection in the x, y-plane.

In many cases it is useful to introduce a third method of describing motion. To this end, we will introduce a means of describing motion that moves with the point mass M along its trajectory. The method of description is based upon the *Serret-Frenet frame* (or *triad*). This triad of basis vectors at a point on the trajectory (Fig. 1.15a) is defined by three orthonormal vectors: $\boldsymbol{e}_t$ in the tangential direction, $\boldsymbol{e}_n$ in the direction of the principal normal, and $\boldsymbol{e}_b$ in the direction of the binormal. The vectors $\boldsymbol{e}_t, \boldsymbol{e}_n$ and $\boldsymbol{e}_b$, in this order, create a right-handed system. The tangent and the principal normal lie in the so-called *osculating plane*. The vector $\boldsymbol{e}_n$ locally points towards the center of curvature C. If M is located at P, the trajectory can be locally approximated by a circle, whose radius ρ (distance $\overline{CP}$) is called the radius of curvature.

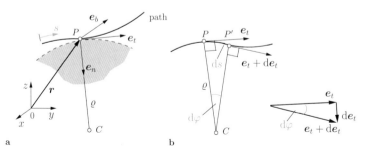

a   b

Fig. 1.15

Using the arc-length $s(t)$, it follows from the expression for the position vector

$$\boldsymbol{r} = \boldsymbol{r}(s(t))$$

that the velocity is given by

$$\boldsymbol{v} = \dot{\boldsymbol{r}} = \frac{\mathrm{d}\boldsymbol{r}}{\mathrm{d}t} = \frac{\mathrm{d}\boldsymbol{r}}{\mathrm{d}s}\frac{\mathrm{d}s}{\mathrm{d}t}.$$

Since $\mathrm{d}\boldsymbol{r}$ points in the direction of the tangent and $|\mathrm{d}\boldsymbol{r}| = \mathrm{d}s$, one has $\mathrm{d}\boldsymbol{r} = \mathrm{d}s\,\boldsymbol{e}_t$. Noting that the *speed* (cf. (1.2)) is

$$v = |\boldsymbol{v}| = \frac{\mathrm{d}s}{\mathrm{d}t} = \dot{s} \tag{1.31}$$

we get

$$\boldsymbol{v} = v\,\boldsymbol{e}_t. \tag{1.32}$$

Differentiation of (1.32) yields the acceleration

$$\boldsymbol{a} = \dot{\boldsymbol{v}} = \dot{v}\,\boldsymbol{e}_t + v\,\dot{\boldsymbol{e}}_t.$$

We determine the time rate of change of the tangent vector, $\dot{\boldsymbol{e}}_t$, analogously to Section 1.1.4. The unit vector $\boldsymbol{e}_t$ changes its direction by an angle $\mathrm{d}\varphi$ between two neighboring points P and P' on the trajectory (Fig. 1.15b). The change $\mathrm{d}\boldsymbol{e}_t$ points towards the center of curvature C and has a magnitude of $1 \cdot \mathrm{d}\varphi$. As the change in arc-length $\mathrm{d}s$ between P and P' can be expressed in terms of the angle $\mathrm{d}\varphi$ and the radius of curvature ρ ($\mathrm{d}s = \rho\,\mathrm{d}\varphi$), it follows that

$$\mathrm{d}\boldsymbol{e}_t = 1 \cdot \mathrm{d}\varphi\,\boldsymbol{e}_n = \frac{\mathrm{d}s}{\rho}\,\boldsymbol{e}_n \quad \rightarrow \quad \dot{\boldsymbol{e}}_t = \frac{\mathrm{d}\boldsymbol{e}_t}{\mathrm{d}t} = \frac{1}{\rho}\frac{\mathrm{d}s}{\mathrm{d}t}\,\boldsymbol{e}_n = \frac{v}{\rho}\,\boldsymbol{e}_n.$$

Substituting back gives an expression for the acceleration vector in terms of the Serret-Frenet frame:

$$\boldsymbol{a} = a_t\,\boldsymbol{e}_t + a_n\,\boldsymbol{e}_n = \dot{v}\,\boldsymbol{e}_t + \frac{v^2}{\rho}\,\boldsymbol{e}_n. \tag{1.33}$$

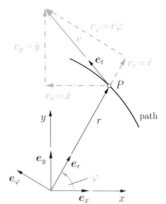

Fig. 1.16

The acceleration is composed of two components: one is in the direction of the tangent to the trajectory, namely the *tangential acceleration* $a_t = \dot{v}$, and one in the direction of the principal normal, namely the *normal acceleration* $a_n = v^2/\rho$. Note that the entire vector lies in the osculating plane.

In the special case of *circular motion*, $\rho = r =$ const, $s = r\varphi$ and $\dot{\varphi} = \omega$. This gives the following velocity and acceleration components:

$$v = \dot{s} = r\omega, \quad a_t = \dot{v} = r\dot{\omega}, \quad a_n = \frac{v^2}{r} = r\omega^2 . \tag{1.34}$$

One can see that this result is consistent with (1.25) when one notes that the direction of the principal normal e_n is opposite to that of e_r .

Between the kinematic variables for rectilinear motion and general three-dimensional motion, one has the following analogous relations:

Rectilinear Motion	*Three-Dimensional Motion*
x	s
$v = \dot{x}$	$v = \dot{s}$
$a = \dot{v} = \ddot{x}$	$a_t = \dot{v} = \ddot{s}$

Thus all the formulae for rectilinear motion from Section 1.1.3 can be used for three-dimensional motion with the appropriate variable substitutions. For example from Table 1.1, if one is given $a_t(v)$, then the arc-length s can be determined from

$$s = s_0 + \int_{v_0}^{v} \frac{\bar{v}\,\mathrm{d}\bar{v}}{a_t(\bar{v})}\ .$$

The Serret-Frenet frame can, of course, also be used to describe planar motions. Fig. 1.16 illustrates three possibilities for constructing an expression for the velocity vector $\boldsymbol{v}$ in the special case of planar motion:

a) Cartesian Coordinates $\boldsymbol{v} = \dot{x}\,\boldsymbol{e}_x + \dot{y}\,\boldsymbol{e}_y,$

b) Polar Coordinates $\boldsymbol{v} = \dot{r}\,\boldsymbol{e}_r + r\dot{\varphi}\,\boldsymbol{e}_\varphi,$

c) Serret-Frenet Frame $\boldsymbol{v} = v\,\boldsymbol{e}_t.$

E1.6   **Example 1.6**   A point mass M moves in the x,y-plane along the trajectory $y = (\alpha/2)\,x^2$ with a constant speed v_0 (Fig. 1.17). Find the magnitude of its acceleration.

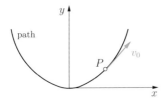

Fig. 1.17

Solution   From (1.33) it follows that a constant speed implies zero tangential acceleration: $a_t = 0$. Thus for this question the normal acceleration a_n and the acceleration magnitude a are one and the same. In order to determine the normal acceleration, we need to find the radius of curvature ρ, which follows for planar curves as

$$\frac{1}{\rho} = \frac{\dfrac{\mathrm{d}^2 y}{\mathrm{d}x^2}}{\left[1 + \left(\dfrac{\mathrm{d}y}{\mathrm{d}x}\right)^2\right]^{3/2}}\ .$$

For the example then,

$$\frac{1}{\rho} = \frac{\alpha}{[1 + (\alpha x)^2]^{3/2}}$$

and we obtain

$$\underline{a} = a_n = \frac{v^2}{\rho} = \frac{\alpha v_0^2}{[1 + (\alpha x)^2]^{3/2}} .$$

As a check on our calculation, we can also solve the problem using Cartesian coordinates. From the expression for the trajectory it follows by differentiation with respect to time that

$$\dot{y} = \alpha x \, \dot{x} .$$

Between the velocity components we also have the constraint that

$$\dot{x}^2 + \dot{y}^2 = v_0^2 .$$

From these two relations we obtain the result that

$$\dot{x}^2 = \frac{v_0^2}{1 + (\alpha x)^2}, \quad \dot{y}^2 = \frac{(\alpha x)^2 \, v_0^2}{1 + (\alpha x)^2} .$$

Differentiating again, gives

$$2\,\dot{x}\,\ddot{x} = -\frac{v_0^2}{[1 + (\alpha x)^2]^2} \, 2\,\alpha^2 x\,\dot{x} \quad \rightarrow \quad \ddot{x} = -\frac{\alpha^2 \, x \, v_0^2}{[1 + (\alpha x)^2]^2} ,$$

$$2\,\dot{y}\,\ddot{y} = \frac{2\,\alpha^2 \, x \, v_0^2}{[1 + (\alpha x)^2]^2}\,\dot{x} \quad \rightarrow \quad \ddot{y} = \frac{\alpha^2 \, x \, v_0^2}{[1 + (\alpha x)^2]^2}\,\frac{\dot{x}}{\dot{y}} = \frac{\alpha \, v_0^2}{[1 + (\alpha x)^2]^2} .$$

With these expressions we obtain our previously derived result

$$a = \sqrt{\ddot{x}^2 + \ddot{y}^2} = \sqrt{\frac{[(\alpha x)^2 + 1]\alpha^2 \, v_0^4}{[1 + (\alpha x)^2]^4}} = \frac{\alpha \, v_0^2}{[1 + (\alpha x)^2]^{3/2}} .$$

In this example, the magnitude of the acceleration is maximal at $x = 0$.

1.2 Kinetics

1.2.1 Newton's Laws

Up to now we have only utilized kinematic quantities (position, velocity, acceleration) to describe motion. We know, however, from experience that motion involves forces in general. We have already studied the concept of forces in detail in Statics (cf. Volume 1). It is now necessary to couple the concept of force to the kinematic quantities. For this purpose, we will restrict our attention in this chapter to the motion of a *point mass*. In a sense, we wish to consider a body whose dimensions have no influence on its motion. Therefore, the body can be represented as a point with a fixed mass m. In what follows, we will usually refer to the body simply as "mass m".

The foundations of kinetics are established in Newton's three laws (1687). They are a summary of all experimental experience and all inferences that can be drawn from them are compatible with common experience. We take these laws – without proof – to be true; i.e. we accept them as axioms.

Newton's 1st Law

> The momentum of a point mass is constant when it is free of external forces.

By *momentum* we mean the *kinetic variable* $\boldsymbol{p}$, which is the product of the mass m and the velocity $\boldsymbol{v}$:

$$\boldsymbol{p} = m\,\boldsymbol{v}\,. \tag{1.35}$$

Momentum is a vector that points in the direction of the velocity. Newton's 1st Law can thus be stated as:

$$\boldsymbol{p} = m\,\boldsymbol{v} = \text{const}\,. \tag{1.36}$$

It says that a point mass executes a uniform rectilinear motion as long as it experiences no net force. Galilei had already formulated this experimental observation in 1638 as the *Law of Inertia* ($v =$ const).

The special case of statics is contained in Newton's 1st Law by considering the case where $v = 0$ (i.e. the body remains still for all times).

Newton's 2nd Law

The time rate of change of the momentum of a mass m is equal to the net external force acting upon it.

In equation form, this law says:

$$\frac{\mathrm{d}p}{\mathrm{d}t} = \frac{\mathrm{d}(m\,v)}{\mathrm{d}t} = F. \tag{1.37}$$

Since it is assumed here that the mass is constant, (1.37) can also be written as

$$m\,\frac{\mathrm{d}v}{\mathrm{d}t} = m\,a = F. \tag{1.38}$$

In our study of the kinetics of a point mass, we will usually employ this form of Newton's 2nd Law, that says in words

Mass $\times$ Acceleration = Force.

The acceleration a has the same direction as the force F.

When the resultant external force is zero, Newton's 1st Law (1.36) follows from (1.37). Thus Newton's 1st Law is simply a special case of Newton's 2nd Law. It is only for historical reasons that we still state them as two separate laws.

The validity of Newton's 2nd Law is subject to two restrictions:

a) The law as stated in (1.38) is valid only in an *inertial reference system*. For the majority of applications, the earth can be considered as an inertial system. How one treats problems when the reference system is non-inertial, i.e. when the reference system is accelerating, will be shown in Chapter 6.

b) In the case where velocities approach the speed of light ($c \approx 300,000$ km/s), one needs to consider the special theory of relativity due to Einstein (1905). This is seldom the case in engineering.

If one releases a body in the vicinity of the earth's surface, it will move under the influence of the *earth's gravitational acceleration* $\boldsymbol{g}$ in the direction of the center of the earth ($g = 9.81$ m/s²). Substituting $\boldsymbol{g}$ into (1.38), we see that during free fall the only force acting on a body is the *weight* $\boldsymbol{W}$, where

$$\boldsymbol{W} = m\,\boldsymbol{g}\,. \tag{1.39}$$

A mass m in the earth's gravitational field has a scalar weight $W = mg$.

If one considers mass, position, and time as the fundamental quantities, then force, according to (1.38), is a derived quantity (cf. Volume 1, Section 1.6). The common unit for force is the *Newton* (1 N = 1 kg ms⁻²).

Newton's 3rd Law
To each force there is an equal and opposite force:

actio = reactio.

The reaction force law (cf. Volume 1, Section 1.5) will make possible the transition from point masses to systems of point masses and then finally to bodies of arbitrary extent.

In addition to these fundamental laws, in the study of kinetics we will also utilize all the basic principles associated with forces (e.g. force parallelograms, section cuts, free-body diagrams) that we know from Statics.

1.2.2 Free Motion, Projectiles

Corresponding to the *three* possibilities for motion in space, a point mass has *three* degrees of freedom. If the motion is not restrained in any direction, one speaks of a *free motion*. This motion is described by the three components of the vector relations (1.38). Considering this, one can pose two types of questions:

a) What are the necessary forces when the trajectory of the motion is known? The solution to this question follows directly from (1.38).

b) What is the motion when the forces are known? This type of question occurs often in engineering situations. From (1.38) the forces directly lead to an expression for the acceleration. If we wish to know the mass's velocity and position we need to integrate once and twice, respectively. For complicated force systems, the integration of the equations of motion can be mathematically quite difficult.

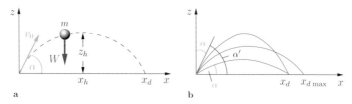

Fig. 1.18

As a simple example, let us consider the case of *projectile motion*. A projectile, modeled as a point mass m, is launched at time $t = 0$ at an angle α with respect to the x-axis with an initial velocity v_0 (Fig. 1.18a). If the air resistance is negligible, then the only force acting on the mass is the weight W in the negative z-direction. In Cartesian coordinates, the equations of motion (1.38) read

$$m\ddot{x} = 0, \quad m\ddot{y} = 0, \quad m\ddot{z} = -W = -mg.$$

Double integration, after cancellation of m, gives:

$$\dot{x} = C_1, \qquad \dot{y} = C_3, \qquad \dot{z} = -gt + C_5,$$

$$x = C_1 t + C_2, \quad y = C_3 t + C_4, \quad z = -g\frac{t^2}{2} + C_5 t + C_6.$$

Out of the three second order differential equations, $3 \cdot 2 = 6$ integration constants appear. These are determined from the 6 initial conditions:

$$\dot{x}(0) = v_0 \cos\alpha \quad \rightarrow \quad C_1 = v_0 \cos\alpha, \quad x(0) = 0 \quad \rightarrow \quad C_2 = 0,$$

$$\dot{y}(0) = 0 \qquad \rightarrow \quad C_3 = 0, \qquad y(0) = 0 \quad \rightarrow \quad C_4 = 0,$$

$$\dot{z}(0) = v_0 \sin\alpha \quad \rightarrow \quad C_5 = v_0 \sin\alpha, \quad z(0) = 0 \quad \rightarrow \quad C_6 = 0.$$

Substituting back provides a parametric solution (parameter t):

$$\dot{x} = v_0 \cos \alpha, \qquad \dot{y} = 0, \quad \dot{z} = -g\,t + v_0 \sin \alpha,$$
$$x = v_0 \cos \alpha \cdot t, \quad y = 0, \quad z = -g\,\frac{t^2}{2} + v_0 \sin \alpha \cdot t. \tag{1.40}$$

One sees that the point mass, which was launched in the x, z-plane, remains in this plane for all time ($y \equiv 0$). In hindsight, this should have been expected as a point mass can not move in the y-direction when there is no force in this direction and the initial velocity $\dot{y}(0)$ is zero. It is also worth noting that the motion is independent of the magnitude of the mass m.

By elimination of the time t from (1.40) one obtains the equation for the curve describing the motion:

$$z(x) = -\frac{g}{2\,v_0^2 \cos^2 \alpha}\, x^2 + \tan \alpha \cdot x. \tag{1.41}$$

This is a quadratic curve, a parabola: the motion of a point mass projectile, launched at an angle, moves on a *parabolic trajectory*.

The projectile distance x_d follows from (1.41) under the condition $z(x_d) = 0$:

$$x_d = \tan \alpha \,\frac{2\,v_0^2 \cos^2 \alpha}{g} = \frac{v_0^2}{g}\,\sin 2\,\alpha. \tag{1.42a}$$

Because $\sin 2\,\alpha = \sin(\pi - 2\,\alpha) = \sin 2\,(\pi/2 - \alpha)$, one obtains the same projectile distance for the same initial velocity v_0 with launch angles α and $\alpha' = \pi/2 - \alpha$ (shallow and steep launch angle, see Fig. 1.18b). The maximum projectile distance occurs when $\alpha = \pi/4$, and results in

$$x_{d\,\text{max}} = \frac{v_0^2}{g}. \tag{1.42b}$$

The projectile time of flight t_d follows by substitution of the projectile distance x_d into (1.40) as

$$t_d = \frac{x_d}{v_0 \cos \alpha} = 2\,\frac{v_0}{g}\,\sin \alpha. \tag{1.43}$$

Comparing a shallow to a steep launch, one sees from (1.43) that the time of flight is larger for a steeper launch angle.

The maximal height z_h of the projectile is found from the condition that at the apex the slope will be zero (the tangent will be horizontal) (Fig. 1.18a):

$$\frac{dz}{dx} = -\frac{g}{v_0^2 \cos^2 \alpha} x + \tan \alpha = 0 \quad \rightarrow \quad x_h = \frac{1}{2} \frac{v_0^2}{g} \sin 2\alpha$$

$$\rightarrow \quad z_h = z(x_h) = \frac{1}{2g} (v_0 \sin \alpha)^2 . \tag{1.44}$$

Due to the symmetry of the trajectory, $x_h = \frac{1}{2} x_d$. Thus the maximal height only depends upon the z-component $\dot{z}(0) = v_0 \sin \alpha$ of the initial velocity.

Example 1.7 A point mass is thrown from a tower (Fig. 1.19a) with an initial velocity v_0 at an angle α with respect to the horizontal. It lands at a distance L from the base of the tower.
a) What is the height H of the tower?
b) How long is the mass in the air?
c) What is the speed of the mass at impact?

E1.7

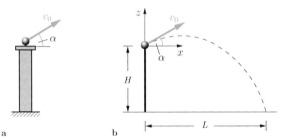

Fig. 1.19   a   b

Solution   a) To start, let us introduce a coordinate system with origin at the top of the tower (Fig. 1.19b). The coordinates of the impact point are then $x = L$ and $z = -H$. Substituting into the equation for the trajectory (1.41) yields

$$H = \frac{g}{2 v_0^2 \cos^2 \alpha} L^2 - L \tan \alpha .$$

b) Since the initial and final locations of the mass do not have the same elevation, we can not determine the time of travel from

(1.43). We need to use the parametric representation (1.40) with $x = L$ and $t = T$, which gives

$$T = \frac{L}{v_0 \cos \alpha}.$$

c) The velocity at impact time $t = T$ has, according to (1.40), two components

$$\dot{x} = v_0 \cos \alpha, \quad \dot{z} = -gT + v_0 \sin \alpha$$

and thus the magnitude

$$\underline{v} = \sqrt{\dot{x}^2 + \dot{z}^2} = \sqrt{v_0^2 \cos^2 \alpha + \left(v_0 \sin \alpha - g \frac{L}{v_0 \cos \alpha} \right)^2}$$

$$= \sqrt{v_0^2 - 2gL \tan \alpha + g^2 \frac{L^2}{v_0^2 \cos^2 \alpha}} = \underline{\sqrt{v_0^2 + 2gH}}.$$

The value of the impact speed is seen to be independent of the initial angle α.

1.2.3 Constrained Motion

When a point mass is restricted to move on a pre-defined surface or curve, then one speaks of a *constrained motion*. In this situation, the number of degrees of freedom is reduced from the *three* degrees of freedom associated with the free motion in space.

The number of degrees of freedom is equal to the number of coordinates necessary to uniquely specify the location of the mass. If the mass moves on a pre-defined surface, then it has *two* degrees of freedom, as a point on a surface requires *two* coordinates for its specification. Any motion orthogonal to the surface is prevented by the constraints. If the point mass is constrained to move on a space-curve, then it has only *one* degree of freedom and its position is specified by a *single* arc-length coordinate s.

In addition to the applied forces $\boldsymbol{F}^{(a)}$ (e.g. the mass's weight), which are independent of the constraints, one also has *constraint forces* $\boldsymbol{F}^{(c)}$, which emanate directly from the constraint surface or curve. These constraint forces are reaction forces that act *orthogonal* to the mass's trajectory. They can be visualized using a free-body diagram which also aids in determining them. With the forces acting on a mass, $\boldsymbol{F}^{(a)}$ and $\boldsymbol{F}^{(c)}$, the dynamical law (1.38) for a constrained mass can be written as

$$m\boldsymbol{a} = \boldsymbol{F}^{(a)} + \boldsymbol{F}^{(c)}. \tag{1.45}$$

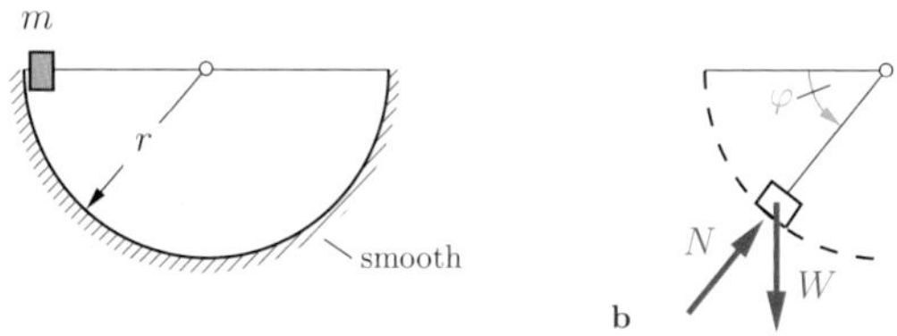

Fig. 1.20

      a   b

    As an example, let us consider the motion of a mass m on a *frictionless* semi-circle of radius r (Fig. 1.20a). The mass is released without an initial velocity at the highest point. As the mass moves on the pre-defined curve (a circle), it has only one degree of freedom. As a coordinate, we choose the angle φ with respect to the horizontal (Fig. 1.20b). Shown in the free-body diagram are the applied force $W = mg$ and the constraint force N. If we describe the motion using a Serret-Frenet frame, then in components (1.45) gives

$$ma_n = F_n^{(a)} + F_n^{(c)}, \quad ma_t = F_t^{(a)}$$

(constraint forces do not have tangential components). In the following, we will indicate the direction of an equation of motion by a properly oriented arrow ($\uparrow$:). For the example, we obtain with $a_n = r\dot{\varphi}^2$ and $a_t = r\ddot{\varphi}$ the equations in the normal and tangential directions:

$$\nearrow: \quad m\,r\dot{\varphi}^2 = N - W\sin\varphi,$$

$$\searrow: \quad m\,r\ddot{\varphi} = W\cos\varphi.$$

These are two equations for the unknowns φ and N. From the second relation using $\ddot{\varphi} = \frac{\mathrm{d}\dot{\varphi}}{\mathrm{d}\varphi}\frac{\mathrm{d}\varphi}{\mathrm{d}t} = \dot{\varphi}\frac{\mathrm{d}\dot{\varphi}}{\mathrm{d}\varphi}$ and separation of variables, one has $\dot{\varphi}\,\mathrm{d}\dot{\varphi} = \frac{g}{r}\cos\varphi\,\mathrm{d}\varphi$ (cf. Section 1.1.3). By integration and using the initial condition, $\dot{\varphi}(\varphi = 0) = 0$, one obtains

$$\frac{\dot{\varphi}^2}{2} = \frac{g}{r} \sin \varphi .$$

The speed is thus $v = r\dot{\varphi} = \sqrt{2\,gr \sin \varphi}$; it takes on its largest value $v_{max} = \sqrt{2\,gr}$ at the lowest point of the trajectory, $\varphi = \pi/2$. If one wishes to determine the path $\varphi(t)$, then by further separation of variables one is led to the integral $\int \dfrac{d\varphi}{\sqrt{\sin \varphi}}$, which is no longer integrable in terms of elementary functions.

The constraint force can be determined from the first equation of motion via substitution of the expression for $\dot{\varphi}^2$:

$$N = mr\,2\,\frac{g}{r} \sin \varphi + W \sin \varphi = 3\,W \sin \varphi .$$

At the lowest point on the trajectory, the constraint force is three times as large as it would be in a static situation.

E1.8

Example 1.8 A horizontal circular plate rotates with constant angular velocity ω_0 (Fig. 1.21a). A point mass m moves in the radial direction within a frictionless slot in the plate. Find the forces acting on the mass under the requirement that it moves with a constant velocity v_0 relative to the plate.

Solution The free-body diagram of the mass is shown in Fig. 1.21b. It shows the two forces acting on the mass: a force F_r, which is necessary to achieve a constant velocity v_0, and a force N_1, which constrains the mass to remain in the slot.

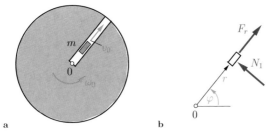

a   b

Fig. 1.21

In polar coordinates, the equations of motion (1.45) read

$$\nearrow: \quad ma_r = F_r, \quad \nwarrow: \quad ma_\varphi = N_1 .$$

From (1.22) with the help of

$$\dot\varphi = \omega_0 \quad \rightarrow \quad \ddot\varphi = 0, \quad \dot r = v_0 \quad \rightarrow \quad \ddot r = 0$$

we can obtain the acceleration components

$$a_r = -r\omega_0^2, \quad a_\varphi = 2\,v_0\,\omega_0 .$$

Substituting into the equations of motion gives the desired forces

$$\underline{\underline{F_r = -m\,r\,\omega_0^2}}, \quad \underline{\underline{N_1 = 2\,m\,v_0\,\omega_0}} .$$

The minus sign in F_r indicates that this force must act inwards. For completeness, it should be mentioned that an additional force N_2 acts orthogonal to the plate; it holds the weight W of the mass in equilibrium: $N_2 = W$.

1.2.4 Resistance/Drag Forces

Resistance forces or *drag forces* hold a special place in the technical theory of mechanics. These are forces that arise due to motion and can be dependent upon the motion itself. Such forces are always tangential to the trajectory and oppose the motion. Common examples include *frictional forces* between bodies and drag forces in aerodynamics.

Let us first consider dry friction. We have already seen *Coulomb's Friction Law*

$$R = \mu N \tag{1.46}$$

in Volume 1. Here, N is the normal force and μ the coefficient of friction. The friction force R is independent of the magnitude of the sliding velocity.

As an example, consider a block of mass m, which, as shown in Fig. 1.22a, slides on a rough plane with inclination angle α. Figure 1.22b shows a free-body diagram with all acting forces: weight W,

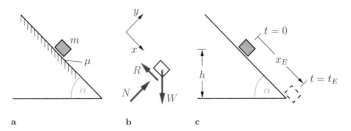

Fig. 1.22

normal force N and friction force R. The equations of motion in the tangential direction x and in the normal direction y are

$$\searrow:\quad m\ddot{x} = mg\sin\alpha - R, \qquad \nearrow:\quad m\ddot{y} = N - mg\cos\alpha\,.$$

With the fact that $\ddot{y} = 0$ (the body is constrained to remain on the plane) it follows from the second equation that $N = mg\cos\alpha$. Substitution of the friction law (1.46) into the first equation gives the acceleration

$$\ddot{x} = g\left(\sin\alpha - \mu\cos\alpha\right) = \text{const}\,.$$

From this expression one can directly determine the position and velocity via integration with respect to time. Assuming the initial conditions $\dot{x}(0) = 0$, $x(0) = 0$, one has

$$\dot{x} = g\left(\sin\alpha - \mu\cos\alpha\right)t\,, \quad x = g\left(\sin\alpha - \mu\cos\alpha\right)\frac{t^2}{2}\,.$$

If the block is released from a height h (see Fig. 1.22c), then it will slide the distance $x_E = h/\sin\alpha$ in the time

$$t_E = t(x_E) = \sqrt{\frac{2\,x_E}{g\left(\sin\alpha - \mu\cos\alpha\right)}} = \sqrt{\frac{2\,h}{g\sin\alpha\left(\sin\alpha - \mu\cos\alpha\right)}}$$

with a final velocity of

$$v_E = \dot{x}(t_E) = g\left(\sin\alpha - \mu\cos\alpha\right)t_E = \sqrt{\frac{2\,gh}{\sin\alpha}\left(\sin\alpha - \mu\cos\alpha\right)}\,.$$

For $\alpha = 90°$ (vertical wall, free fall), $N = 0$. In this case, there will be no frictional force and the final velocity v_E will be the same as the impact velocity $v_E = \sqrt{2\,gh}$ from Section 1.1.3.

For the motion of a solid body in a liquid or gaseous medium, one also has resistance forces, which are normally known as *drag forces*. Out of the multitude of drag forces that one can determine experimentally, we will focus our attention on two idealized cases.

At low velocities, flows are laminar. The drag force F_d in this situation is proportional to the velocity:

$$F_d = kv. \tag{1.47a}$$

Here, the constant k depends upon the shape of the body and the viscosity η of the fluid. George Gabriel Stokes (1819-1903) determined in 1854 the relation for the drag force on a sphere of radius r in a fluid with velocity v (or on a sphere which moves with velocity v in a stationary fluid) as

$$F_d = 6\,\pi\,\eta\,r\,v\,. \tag{1.47b}$$

A linear relation between velocity and resistance force will also be seen to be a common assumption in the analysis of damped oscillations (Chapter 5).

At larger velocities, the flow becomes turbulent. In this case, the drag force can be estimated as

$$F_d = kv^2\,, \tag{1.48a}$$

where the constant k depends upon the geometry of the body and the density of the fluid. This relation is often written in the form

$$F_d = C_d\,\frac{\rho}{2}\,A\,v^2\,. \tag{1.48b}$$

Here, A is the projected area of the body onto the plane orthogonal to the direction of the flow and the drag coefficient C_d accounts for all other parameters. For example, for modern automobiles, it has a value smaller than 0.3.

As an illustrative example, let us consider the velocity of a body during free fall with drag. A body with weight W is released from an arbitrary height with zero initial velocity. The drag force on the body is assumed to be given by (1.48a). Using the notation introduced in Fig. 1.23a, the equation of motion reads

$$\downarrow: \quad m\ddot{x} = W - F_d = mg - k\dot{x}^2 \,.$$

Introducing for convenience the constant $\kappa^2 = mg/k$, one has

$$\ddot{x} = g\left(1 - \frac{\dot{x}^2}{\kappa^2}\right) \,. \tag{a}$$

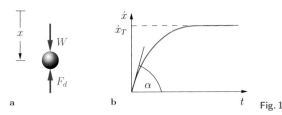

a   b   t   Fig. 1.23

By separation of variables and integration, we find

$$\frac{\mathrm{d}\dot{x}}{g\left(1 - \dfrac{\dot{x}^2}{\kappa^2}\right)} = \mathrm{d}t \quad \rightarrow \quad t = \frac{\kappa}{g}\tanh^{-1}\frac{\dot{x}}{\kappa} + C \,.$$

Since at time $t = 0$ the velocity $\dot{x}(0) = 0$, we note that $C = 0$.
Solving for $\dot{x}$, gives the velocity function

$$\dot{x} = \kappa\tanh\frac{g\,t}{\kappa} \,.$$

For increasing t, the velocity asymptotically approaches the limiting value $\dot{x}_T = \kappa$, since the hyperbolic tangent approaches
unity for large arguments. The motion becomes uniform. We can
also determine the terminal velocity $\dot{x}_T$ directly from (a) under
the condition $\ddot{x} = 0$; this allows us to simply read-off the result
$\dot{x}_T = \kappa$.

Figure 1.23b shows the time history of the velocity. At the start
of the motion, the velocity is zero and thus according to (a) the
acceleration $\ddot{x} = g$: the initial slope $\mathrm{d}\dot{x}/\mathrm{d}t$ (equivalently the angle
α) is determined by the gravitational acceleration. For increasing t
the velocity $\dot{x}$ approaches the limiting value $\dot{x}_T$, which it achieves
only in infinite time.

Example 1.9 A conveyer belt moves with constant velocity $v_B = 3 \, \text{m/s}$. At time $t = 0$ (see Fig. 1.24a) a crate with weight $W = mg$ and horizontal velocity $v_0 = 0.5$ m/s is placed on the belt at position A; the coefficient of friction is $\mu = 0.2$.

E1.9

How long does the crate slide with respect to the belt? What is the position of the crate once it stops sliding with respect to the initial position it had on the belt?

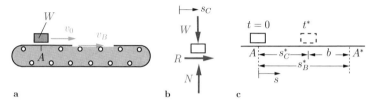

Fig. 1.24

Solution Since $v_B > v_0$, a frictional force R acts on the crate to the right (Fig. 1.24b). From the equations of motion for the crate

$$\rightarrow: \quad m\ddot{s}_C = R \,,$$

it follows with $N = mg$ and the friction law

$$R = \mu N = \mu mg \,,$$

that the acceleration is given by

$$\ddot{s}_C = \mu g.$$

Using the initial conditions $\dot{s}_C(0) = v_0$ and $s_C(0) = 0$, we find via integration

$$\dot{s}_C = v_C = \mu g t + v_0, \quad s_C = \mu g \frac{t^2}{2} + v_0 t \,.$$

The sliding ends at time t^*, when the crate's velocity equals the conveyer belt's velocity v_B:

$$v_C = v_B \quad \rightarrow \quad \mu g t^* + v_0 = v_B.$$

Using the given values, it follows from this result that the sliding process occurs for a time

$$\underline{\underline{t^*}} = \frac{v_B - v_0}{\mu g} = \frac{3 - 0.5}{0.2 \cdot 9.81} = \underline{\underline{1.27 \text{ s}}} \,.$$

At time t^*, the crate has moved a distance

$$s_C^* = s_C(t^*) = \mu g \, \frac{t^{*2}}{2} + v_0 \, t^*$$

$$= 0.2 \cdot 9.81 \cdot \frac{1.27^2}{2} + 0.5 \cdot 1.27 = 2.2 \text{ m} \,.$$

In the same time, the location A on the conveyer belt has moved a distance

$$s_B^* = s_B(t^*) = v_B \, t^* = 3 \cdot 1.27 = 3.8 \text{ m}$$

to the position A^* (Fig. 1.24c). The distance b between the crate and its initial location on the conveyer belt is thus

$$\underline{\underline{b}} = s_B^* - s_C^* = 3.8 - 2.2 = \underline{\underline{1.6 \text{ m}}} \,.$$

E1.10

Example 1.10 A sphere (mass m, radius r) is dropped in a container filled with liquid (Fig. 1.25a).

Determine the motion under the assumptions that the drag force is given by Stoke's Law and that the buoyancy force is negligible.

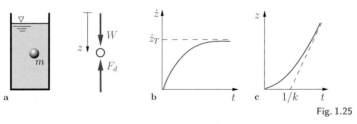

Fig. 1.25

Solution   As shown in Fig. 1.25b, let us take the positive z-coordinate as pointing down. According to (1.47b), the motion can be described by the equation

$$\downarrow: \quad m\ddot{z} = W - F_d = mg - 6\pi\eta r\dot{z}\,.$$

Using the shorthand, $\dfrac{6\pi\eta r}{m} = k$, gives

$$\ddot{z} = g - k\dot{z}\,.$$

Separation of variables and integration lead to

$$\frac{\mathrm{d}\dot{z}}{g - k\dot{z}} = \mathrm{d}t \quad \rightarrow \quad -\frac{1}{k}\ln\left(1 - \frac{k\dot{z}}{g}\right) = t + C_1\,.$$

If we assume the initial condition $\dot{z}(0) = 0$, then $C_1 = 0$. Inverting the above relation, one obtains the expression for the velocity:

$$\dot{z} = \frac{g}{k}\left(1 - \mathrm{e}^{-kt}\right)\,.$$

With an additional integration, one can determine the position of the mass:

$$z = \frac{g}{k}\left(t + \frac{1}{k}\mathrm{e}^{-kt} + C_2\right)\,.$$

Using the initial condition $z(0) = 0$ gives $C_2 = -1/k$ and thus

$$z = \frac{g}{k}\left[t - \frac{1}{k}\left(1 - \mathrm{e}^{-kt}\right)\right]\,.$$

As $kt \rightarrow \infty$ (the long-time limit), $\dot{z}$ tends towards the limiting value

$$\dot{z}_T = \frac{g}{k} = \mathrm{const}\,.$$

In this limit, the position becomes a linear function of time. For large k, e.g. for large viscosity η, this limit is practically instantly reached. A measurement of the constant velocity $\dot{z}_T$ can then serve as a means for determining η. Figure 1.25c shows the time history of the velocity and position.

1.2.5 Impulse Law and Linear Momentum, Impact

If one integrates Newton's 2nd Law

$$\frac{\mathrm{d}}{\mathrm{d}t}(m\,\boldsymbol{v}) = \boldsymbol{F}$$

with respect to time, one obtains the *Impulse Law* or *principle of impulse and linear momentum*

$$m\,\boldsymbol{v} - m\,\boldsymbol{v}_0 = \int_{t_0}^{t} \boldsymbol{F}\,\mathrm{d}\bar{t}. \tag{1.49}$$

Thus, the change in the momentum $\boldsymbol{p} = m\,\boldsymbol{v}$ between time t_0 and an arbitrary time t is equal to the time integral of the force. If $\boldsymbol{F}$ is zero over this time interval, then the momentum is unchanged (conservation of linear momentum):

$$\boldsymbol{p} = m\,\boldsymbol{v} = m\,\boldsymbol{v}_0 = \text{const} .$$

The Impulse Law is often used to study impact processes. An *impact* is defined by a large force that acts over a very short time span (impact duration t_i). In this situation, the mass experiences a sudden change in velocity but the change in its position is negligible. The precise time variation of $\boldsymbol{F}$ during the impact is usually unknown. In order to determine the velocity after an impact, we introduce the concept of *linear impulse* $\hat{\boldsymbol{F}}$, the time integral of the impact forces over the impact interval:

$$\hat{\boldsymbol{F}} = \int_{0}^{t_i} \boldsymbol{F}\,\mathrm{d}t . \tag{1.50}$$

Thus from (1.49) for an impact process

$$m\,\boldsymbol{v} - m\,\boldsymbol{v}_0 = \hat{\boldsymbol{F}} . \tag{1.51}$$

Let us consider now a body modelled as a point mass, which as shown in Fig. 1.26a, obliquely strikes a wall. In what follows, we will denote the velocity before the impact as $\boldsymbol{v}$ and after impact as $\bar{\boldsymbol{v}}$. With the given coordinate system, (1.51) reads

$$\rightarrow: \quad m\,\bar{v}_x - m\,v_x = \hat{F}_x, \quad \uparrow: \quad m\,\bar{v}_y - m\,v_y = \hat{F}_y\,, \qquad (1.52)$$

where the arrow (e.g. $\rightarrow:$) shows in which direction the Impulse Law has been written. From Fig. 1.26a, we can see that

$$v_x = -v\cos\alpha, \quad v_y = v\sin\alpha,$$

$$\bar{v}_x = \bar{v}\cos\bar{\alpha}, \qquad \bar{v}_y = \bar{v}\sin\bar{\alpha}\,.$$

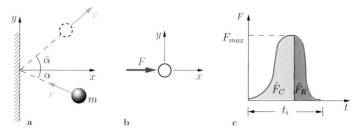

Fig. 1.26

Let us further assume that the wall is *frictionless*. Then, no forces can be generated in the y-direction and with $\hat{F}_y = 0$ it follows from (1.52) that

$$\bar{v}_y = v_y\,. \qquad (1.53)$$

The velocity component in the y-direction does not change during frictionless impact.

To determine the x-component of the velocity, we will decompose the impact interval into two parts: the *compression phase*, in which the body represented by the point mass is compressed and its center of mass comes nearer to the wall, and the *restitution phase*, during which the center of mass of the body moves away from the wall. The force $F_x = F$, acting on the body during the impact process (Fig. 1.26b), increases during the compression phase to a maximal value $F_{\max}$ and decreases during the restitution phase back to zero (Fig. 1.26c). We now write the Impulse

Law in the x-direction for the two phases:

$$\text{Compression phase:} \quad m \cdot 0 - m\,v_x = \hat{F}_C,$$
$$\text{Restitution phase:} \quad m\,\bar{v}_x - m \cdot 0 = \hat{F}_R \tag{1.54}$$

(at the moment of maximal force the velocity is zero). The two equations (1.54) contain three unknowns: the velocity $\bar{v}_x$ and the two linear impulses $\hat{F}_C$ and $\hat{F}_R$. To complete the system of equations, we need to make an assumption about the impact process itself. There are three basic cases (or models) that are commonly applied:

a) *Ideal-elastic impact*

We assume that the deformations and forces in the compression phase and the restitution phase are mirror images of each other. In this case, the linear impulses in both phases are assumed to be equal. From $\hat{F}_R = \hat{F}_C$, it follows that

$$m\,\bar{v}_x = -\,m\,v_x \quad \rightarrow \quad \bar{v}_x = -\,v_x$$

and from (1.53) it follows that

$$\bar{v} = v \quad \text{and} \quad \bar{\alpha} = \alpha\,.$$

In an ideal-elastic impact (Fig. 1.27a), the incident and ricochet angles and velocities are equal (cf. Reflection Law of Optics).

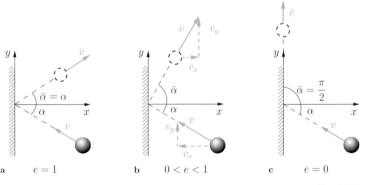

a $e = 1$ b $0 < e < 1$ c $e = 0$

Fig. 1.27

b) *Ideal-plastic impact*

The deformation experienced by the body during the compression phase is permanent; we assume that the linear impulse of the restitution phase vanishes ($\hat{F}_R = 0$). Then according to (1.54), it follows that

$$\bar{v}_x = \bar{v}\cos\bar{\alpha} = 0 \quad \rightarrow \quad \bar{\alpha} = \frac{\pi}{2}\,.$$

The body is observed to slide with velocity $\bar{v} = \bar{v}_y = v_y = v\sin\alpha$ along the frictionless wall after impact (Fig. 1.27c).

c) *Partially elastic impact*

A real body responds in a fashion intermediate to the two limiting cases of ideal-elastic and ideal-plastic impact. To model a real body, one assumes a simple proportionality relation between the linear impulses of compression and restitution. The proportionality constant e is known as the *coefficient of restitution*:

$$\hat{F}_R = e\,\hat{F}_C\,. \tag{1.55}$$

In the limiting cases of ideal-elastic impact and ideal-plastic impact, $e = 1$ and $e = 0$, respectively. For partially elastic impact, the coefficient of restitution lies between these two limits; i.e.,

$$0 \leqq e \leqq 1\,. \tag{1.56}$$

Substitution of (1.54) into (1.55) gives

$$m\,\bar{v}_x = e\,(-m\,v_x) \quad \rightarrow \quad \bar{v}_x = -\,e\,v_x\,. \tag{1.57}$$

According to Fig. 1.27b this implies that $\tan\bar{\alpha} = \dfrac{\bar{v}_y}{\bar{v}_x} = \dfrac{v_y}{-e v_x} = \dfrac{1}{e}\tan\alpha$. Since $e < 1$ for partially elastic impact, $\tan\bar{\alpha} > \tan\alpha$ and thus $\bar{\alpha} > \alpha$.

Using equation (1.57), one can also construct a working definition of the coefficient of restitution in terms of the velocity components orthogonal to the wall before and after the impact:

$$e = -\,\frac{\bar{v}_x}{v_x}\,. \tag{1.58}$$

The minus sign appears because the two velocities are taken as positive in the same direction. In the example, v_x is negative and thus e is positive.

The coefficient of restitution e can be determined experimentally. If one drops a mass from a height h_1 onto a horizontal surface, then the (downward) incident velocity is according to Section 1.1.3

$$v = \sqrt{2\,g\,h_1}\,.$$

After impact, if the body has a (upwards) rebound velocity $\bar{v}$, then it will reach a height

$$h_2 = \frac{\bar{v}^2}{2\,g} \quad \rightarrow \quad \bar{v} = \sqrt{2\,g\,h_2}\,.$$

Taking proper account of the signs, it follows from (1.58) that

$$e = -\frac{\bar{v}}{v} = \frac{\sqrt{2\,g\,h_2}}{\sqrt{2\,g\,h_1}} \quad \rightarrow \quad e = \sqrt{\frac{h_2}{h_1}}\,. \tag{1.59}$$

Using this method, the coefficient of restitution can be directly calculated from the heights before and after the impact. For an ideal-elastic impact, $h_2 = h_1$ since $e = 1$; for an ideal-plastic impact $h_2 = 0$ since $e = 0$.

E1.11   **Example 1.11** A man (weight $W_1 = m_1\,g$) stands on the runners of a sled (weight $W_2 = m_2\,g$) and kicks-off the ground at uniformly separated times (separation Δt), such that the initially stationary sled begins to move (Fig. 1.28a). The friction coefficient μ between the ground and the sled is given. To simplify the situation it will be assumed that each kick occurs over a short time t_s ($t_s \ll \Delta t$) and with a *constant* horizontal force P.

How large is the velocity v directly after the n-th kick?

Solution   We can model the system (sled and man) as a point mass (Fig. 1.28b). A horizontally oriented force P acts on the mass during each impact of duration t_s. Further, a friction force $R = \mu N$ acts on the mass over all times. Up through the n-th kick, a time $T = (n-1)\Delta t$ passes. Assuming an initial velocity $v_0 = 0$ at time $t_0 = 0$, one computes from the Impulse Law (1.49)

Fig. 1.28

for the total mass

$$\leftarrow: \quad (m_1 + m_2)\, v = \int_0^T F\,\mathrm{d}t = n \int_0^{t_s} P\,\mathrm{d}t - \int_0^T \mu(W_1 + W_2)\,\mathrm{d}t\ .$$

The sought velocity after the n-th kick is then

$$v = \frac{n\,P\,t_s}{m_1 + m_2} - \mu\,g\,(n-1)\,\Delta t\ .$$

Example 1.12 A hockey puck strikes a frictionless wall with velocity v at an angle $\alpha = 45°$ and bounces off at an angle $\beta = 30°$ (Fig. 1.29).

E1.12

Find the ricochet velocity $\bar{v}$ and the coefficient of restitution.

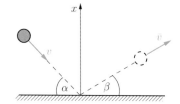

Fig. 1.29

Solution For a *frictionless* wall the momentum parallel to the wall must be conserved:

$$\rightarrow: \quad m\,\bar{v}\,\cos\beta = m\,v\,\cos\alpha\ .$$

From this, it follows that the velocity after the impact is

$$\underline{\underline{\bar{v}}} = v\,\frac{\cos\alpha}{\cos\beta} = \sqrt{\frac{2}{3}}\,v\,.$$

Using the velocity components orthogonal to the wall,

$$v_x = -\,v\sin\alpha\,,\quad \bar{v}_x = \bar{v}\sin\beta\,,$$

one can determine the coefficient of restitution e from (1.58):

$$\underline{\underline{e}} = -\,\frac{\bar{v}\sin\beta}{-\,v\sin\alpha} = \sqrt{\frac{2}{3}}\,\frac{\frac{1}{2}}{\frac{1}{2}\sqrt{2}} = \underline{\underline{\frac{\sqrt{3}}{3}}}\,.$$

1.2.6 Angular Momentum Theorem

In the study of statics (cf. Volume 1), we introduced for the moment of a force relative to a point 0 the moment vector

$$\boldsymbol{M}^{(0)} = \boldsymbol{r} \times \boldsymbol{F}\,. \tag{1.60}$$

An analogous kinetic variable is the *moment of momentum* $\boldsymbol{L}^{(0)}$. It is defined as the vector product of $\boldsymbol{r}$ and $\boldsymbol{p}$:

$$\boldsymbol{L}^{(0)} = \boldsymbol{r} \times \boldsymbol{p} = \boldsymbol{r} \times m\,\boldsymbol{v}\,. \tag{1.61}$$

The vector $\boldsymbol{L}^{(0)}$ is also known as the *angular momentum vector*. It is orthogonal to the plane containing the position vector $\boldsymbol{r}$ (from a fixed point 0 to the moving point mass) and the velocity vector $\boldsymbol{v}$ (Fig. 1.30a). Its magnitude is given by the product of the orthogonal projection $r_\perp$ (to the velocity) and the linear momentum magnitude mv as $L^{(0)} = r_\perp mv$.

We wish now to determine a relation between angular momentum and moment. To this effect, we form the vector product of the position vector with Newton's 2nd Law (1.38):

$$\boldsymbol{r} \times \left(m\,\frac{\mathrm{d}\boldsymbol{v}}{\mathrm{d}t} \right) = \boldsymbol{r} \times \boldsymbol{F}\,. \tag{1.62}$$

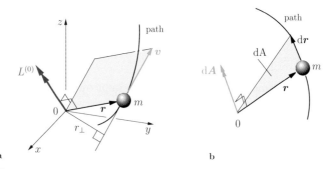

Fig. 1.30

The right-hand side is the moment as defined in (1.60). The left-hand side can be re-written with the help of the identity

$$\frac{\mathrm{d}}{\mathrm{d}t}(\boldsymbol{r} \times m\,\boldsymbol{v}) = \dot{\boldsymbol{r}} \times m\,\boldsymbol{v} + \boldsymbol{r} \times m\,\frac{\mathrm{d}\boldsymbol{v}}{\mathrm{d}t} \;,$$

which follows directly from the chain rule of differentiation. The first term in this identity on the right-hand side vanishes since $\dot{\boldsymbol{r}} = \boldsymbol{v}$. Thus,

$$\boldsymbol{r} \times m\,\frac{\mathrm{d}\boldsymbol{v}}{\mathrm{d}t} = \frac{\mathrm{d}}{\mathrm{d}t}(\boldsymbol{r} \times m\,\boldsymbol{v}) = \frac{\mathrm{d}\boldsymbol{L}^{(0)}}{\mathrm{d}t} \;.$$

This allows us to write (1.62) as

$$\frac{\mathrm{d}\boldsymbol{L}^{(0)}}{\mathrm{d}t} = \boldsymbol{M}^{(0)} \;. \tag{1.63}$$

This is the *angular momentum theorem*: the time rate of change of the angular momentum of a point mass relative to a *fixed* arbitrary point 0 is equal to the moment of the force acting on the point mass relative to 0.

If the moment $\boldsymbol{M}^{(0)}$ is zero, then the angular momentum remains constant (*conservation of angular momentum*):

$$\boldsymbol{L}^{(0)} = \boldsymbol{r} \times m\,\boldsymbol{v} = \text{const} \;.$$

A visual interpretation of angular momentum can be found by considering Fig. 1.30b. In a time increment $\mathrm{d}t$ the position vector

r sweeps out an area with magnitude $dA = \frac{1}{2}|r \times dr|$. If we introduce the corresponding vectorial quantity

$$dA = \frac{1}{2}(r \times dr) = \frac{1}{2}(r \times v\,dt) \; ,$$

then the time rate of change of the swept out area is

$$\frac{dA}{dt} = \frac{1}{2}(r \times v) \; .$$

Substituting back into (1.61) results in

$$L^{(0)} = 2\,m\,\frac{dA}{dt} \; . \tag{1.64}$$

Thus, the angular momentum is proportional to the time rate of change of the swept out area.

If for a given motion, the force points continuously towards a fixed point 0, then the moment relative to 0 will vanish. In this case, the angular momentum and the time rate of change of the swept out area, according to (1.64), will be constant. For planetary motion, this result is known as Kepler's 2nd Law (or the Law of Equal Areas): the ray from the sun to a planet sweeps out equal areas in equal times (cf. Sec. 1.1.4).

If a point mass moves only in the x, y-plane (Fig. 1.31), then the angular momentum vector and the moment vector only have non-zero z-components. Thus in the angular momentum theorem (1.63) only one component remains:

$$\frac{dL_z^{(0)}}{dt} = M_z^{(0)} \; . \tag{1.65}$$

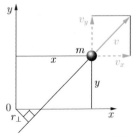

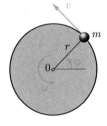

Abb. 1.31   Abb. 1.32

In this case we can drop the subscript z for simplicity. The angular momentum – analogous to the moment – can either be expressed using the orthogonal (with respect to the velocity) distance $r_\perp$ from the reference point or via components v_x and v_y and the explicit expression for the vector product:

$$L^{(0)} = r_\perp\, m\, v \quad \text{or} \quad L^{(0)} = m(x\, v_y - y\, v_x)\,. \tag{1.66}$$

In the special case of circular motion (Fig. 1.32), we find using $v = r\,\omega$ that the angular momentum is

$$L^{(0)} = m\, r\, v = m\, r^2\omega\,.$$

Let us introduce the symbol $\Theta^{(0)}$ for the term $m r^2$. This is known as the *mass moment of inertia* or simply *moment of inertia*. Thus the angular momentum can be expressed as $L^{(0)} = \Theta^{(0)}\omega$, and using the connection $\omega = \dot\varphi$ allows the angular momentum theorem (1.65) to be written as

$$\Theta^{(0)}\ddot\varphi = M^{(0)}\,. \tag{1.67}$$

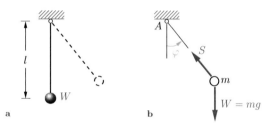

Fig. 1.33

    a   b

As an example, let us consider the motion of a pendulum as shown in Fig. 1.33a. There are two forces acting on m: a support force S (pointing towards A) and the gravitational force $W = mg$ (Fig. 1.33b). Introducing a positive rotation angle φ as indicated allows us to express the angular momentum and moment relative to the fixed point A as

$$L^{(A)} = l\, m\, v = l\, m\, l\dot\varphi = ml^2\,\dot\varphi, \quad M^{(A)} = -\, mg\, l \sin\varphi\,.$$

The angular momentum theorem (1.65) then furnishes an expression for the equation of motion:

$$ml^2 \ddot{\varphi} = -mgl\sin\varphi \quad \rightarrow \quad \ddot{\varphi} + \frac{g}{l}\sin\varphi = 0.$$

For small angles ($\sin\varphi \approx \varphi$), this can be written as $\ddot{\varphi} + \frac{g}{l}\varphi = 0$ (simple harmonic oscillation of an *ideal pendulum*, cf. Chapter 5). With $\Theta^{(A)} = ml^2$ one can also determine the equation of motion from (1.67).

E1.13

Example 1.13 A mass m executes a circular motion with an angular velocity ω_0 on a frictionless horizontal plane. The mass is held at a radius r_0 with a string (Fig. 1.34a, b). The string is threaded through a hole A at the center of the plane.

a) If the string is pulled, such that the mass moves on a circular trajectory with radius r, what will the angular velocity be?

b) What is the corresponding change of the force in the string?

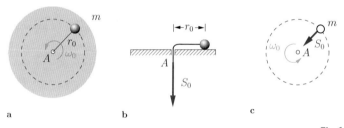

Fig. 1.34

Solution a) Let us denote the force in the string in the initial state as S_0 (Fig. 1.34c); this force has no moment about A. Thus according to (1.65), the angular momentum must be constant during the motion and will remain so even after the string has been pulled. The initial angular momentum about A is

$$L_0^{(A)} = r_0(m\,r_0\,\omega_0) = m\,r_0^2\,\omega_0$$

and after pulling the string

$$L^{(A)} = r(m\,r\omega) = m\,r^2\,\omega\,.$$

Setting the two expressions equal gives

$$\omega = \left(\frac{r_0}{r}\right)^2 \omega_0 .$$

The angular velocity is inversely proportional to the square of the radius of the trajectory.

b) Noting that the centripetal acceleration $a_n = v^2/r = r\omega^2$, the equation of motion in the initial state (Fig. 1.34c) gives

$$\diagup : \quad m\, a_n = S_0 \quad \rightarrow \quad S_0 = m\, r_0\, \omega_0^2 .$$

In the same manner, for the final state, we have

$$S = m\, r\, \omega^2 = m\, r \left(\frac{r_0}{r}\right)^4 \omega_0^2 = \left(\frac{r_0}{r}\right)^3 m\, r_0\, \omega_0^2 = \left(\frac{r_0}{r}\right)^3 S_0 .$$

The force in the string is inversely proportional to the cube of the radius of the trajectory.

1.2.7 Work-Energy Theorem, Potential Energy, Conservation of Energy

If we form the scalar product of Newton's 2nd Law (1.38) with $\mathrm{d}\boldsymbol{r}$, then we find

$$m\, \frac{\mathrm{d}\boldsymbol{v}}{\mathrm{d}t} \cdot \mathrm{d}\boldsymbol{r} = \boldsymbol{F} \cdot \mathrm{d}\boldsymbol{r} .$$

Substituting $\mathrm{d}\boldsymbol{r} = \boldsymbol{v}\, \mathrm{d}t$ and integrating between two points $\boldsymbol{r}_0$ and $\boldsymbol{r}_1$ on a mass's trajectory, gives

$$\int_{\boldsymbol{v}_0}^{\boldsymbol{v}_1} m\, \boldsymbol{v} \cdot \mathrm{d}\boldsymbol{v} = \int_{\boldsymbol{r}_0}^{\boldsymbol{r}_1} \boldsymbol{F} \cdot \mathrm{d}\boldsymbol{r} \quad \rightarrow \quad \frac{m\boldsymbol{v}_1^2}{2} - \frac{m\boldsymbol{v}_0^2}{2} = \int_{\boldsymbol{r}_0}^{\boldsymbol{r}_1} \boldsymbol{F} \cdot \mathrm{d}\boldsymbol{r} , \quad (1.68)$$

where $\boldsymbol{v}_0$ and $\boldsymbol{v}_1$ are the mass's velocity at these two points. The right-hand side expresses the work U done by the force $\boldsymbol{F}$ (cf. Volume 1, Chapter 8). The scalar quantity $\frac{1}{2}\, m\, \boldsymbol{v}^2 = \frac{1}{2}\, m\, v^2$ is called the *kinetic energy* T:

$$T = \frac{mv^2}{2} . \qquad\qquad (1.69)$$

Using this, we obtain from (1.68) the *work-energy theorem*

$$T_1 - T_0 = U \,. \tag{1.70}$$

The work done by the forces acting on a mass between two points on its trajectory is equal to the change in the mass's kinetic energy.

Just as with work U, the dimensions of kinetic energy T are force × distance. In many applications it is expressed using the unit *Joule*, 1 J = 1 Nm, (James Prescott Joule, 1818-1889).

The forces acting on a point mass are composed of externally applied forces $\boldsymbol{F}^{(a)}$ and constraint forces $\boldsymbol{F}^{(c)}$. Since the constraint forces are orthogonal to the trajectory of a mass, they do no work. Thus, the work integral is given by

$$U = \int_{\boldsymbol{r}_0}^{\boldsymbol{r}_1} \boldsymbol{F}^{(a)} \cdot \mathrm{d}\boldsymbol{r} \,. \tag{1.71}$$

As an example application consider a block sliding down a rough inclined plane (Fig. 1.35a). The block is acted upon by an applied gravitational force $W = mg$ and a friction force $R = \mu N$, in addition to the constraint force N (Fig. 1.35b).

In moving from position ⓪ to position ① the gravitational force and the friction force do work:

$$U_W = mg \sin \alpha \, x, \quad U_R = -Rx = -\mu N \, x = -\mu mg \cos \alpha \, x$$

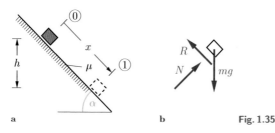

a   b   Fig. 1.35

(the constraint force N does no work). If the block is released from position ⓪ with zero initial velocity, then the work-energy theorem (1.70) tells us that

$$m \frac{v_1^2}{2} = mg \sin \alpha \, x - mg \, \mu \cos \alpha \, x \,.$$

Introducing the height $h = x \sin \alpha$ allows us to express the velocity in position ① as

$$v_1 = \sqrt{2 \, gh(1 - \mu \cot \alpha)} \,.$$

The result shows that the motion is only possible if $\mu \cot \alpha < 1$, i.e. $\mu < \tan \alpha$.

The work done per unit time dU/dt is known as the *power* P. Writing $dU = \boldsymbol{F} \cdot d\boldsymbol{r}$ gives

$$P = \boldsymbol{F} \cdot \boldsymbol{v} \,. \tag{1.72}$$

The most common unit of power is the *Watt* (James Watt, 1736-1819):

$$1 \text{ W} = 1 \frac{\text{Nm}}{\text{s}} \,.$$

Another common unit for power is horsepower (hp); its relation to Watt is given as

$$1 \text{ hp} = 0.735 \text{ kW}, \quad 1 \text{ kW} = 1.36 \text{ hp} \,.$$

As with work done, the power due to constraint forces $\boldsymbol{F}^{(c)}$ is zero, as they are orthogonal to a point mass's velocity $\boldsymbol{v}$.

In all machines, energy is lost due to frictional effects in supports and guides. Thus a part of all input or applied work is simply lost. One expresses the relationship between output or usable energy U_O and input or applied energy U_A as the *efficiency* η:

$$\eta = \frac{U_O}{U_A} \,. \tag{1.73}$$

As a per unit time quantity, one has the instantaneous efficiency as the ratio of the corresponding powers:

$$\eta = \frac{P_O}{P_A} \,. \tag{1.74}$$

Due to the ever present losses $\eta < 1$.

As an example application let us compute the output drive force F of an automobile with a motor rated at $P_A = 30$ kW that is traveling at a velocity of $v = 60$ km/h on a flat road. We will assume an efficiency $\eta = 0.8$. The output or drive power is $P_O = Fv$, thus from (1.74)

$$\eta = \frac{Fv}{P_A} \quad \rightarrow \quad F = \frac{P_A \, \eta}{v} = \frac{30 \cdot 0.8}{60/3.6} = 1.44 \text{ kN}\,.$$

The work-energy theorem (1.70) takes on a particularly simple form when the applied forces emanate from a potential. Such forces are known as *conservative forces*. Forces of this type are characterized by the fact that the work they perform between two fixed points in space ⓪ and ① (Fig. 1.36) is independent of the path taken between these points (cf. Volume 1, Chapter 8). Writing $\boldsymbol{F} = F_x \, \boldsymbol{e}_x + F_y \, \boldsymbol{e}_y + F_z \, \boldsymbol{e}_z$ and $\mathrm{d}\boldsymbol{r} = \mathrm{d}x \, \boldsymbol{e}_x + \mathrm{d}y \, \boldsymbol{e}_y + \mathrm{d}z \, \boldsymbol{e}_z$ the work done between the two points is given by

$$U = \int_{⓪}^{①} \boldsymbol{F} \cdot \mathrm{d}\boldsymbol{r} = \int_{⓪}^{①} \{ F_x \, \mathrm{d}x + F_y \, \mathrm{d}y + F_z \, \mathrm{d}z \} \,. \tag{1.75}$$

The integral is *path independent* only when the integrand is an *exact differential* (total differential), which we will denote as $-\mathrm{d}V$:

$$-\, \mathrm{d}V = F_x \, \mathrm{d}x + F_y \, \mathrm{d}y + F_z \, \mathrm{d}z\,. \tag{1.76}$$

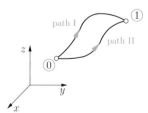

Fig. 1.36

The function $V(x, y, z)$ is called the *potential energy* or *potential* of the force $\boldsymbol{F}$; the minus sign in the definition has been introduced due to its utility with respect to physical interpretation as we will see later.

A comparison of the exact differential

$$dV = \frac{\partial V}{\partial x}\,dx + \frac{\partial V}{\partial y}\,dy + \frac{\partial V}{\partial z}\,dz$$

with (1.76) gives the relations

$$F_x = -\frac{\partial V}{\partial x}, \quad F_y = -\frac{\partial V}{\partial y}, \quad F_z = -\frac{\partial V}{\partial z}. \tag{1.77}$$

If we introduce the *gradient*

$$\operatorname{grad} V = \frac{\partial V}{\partial x}\,\boldsymbol{e}_x + \frac{\partial V}{\partial y}\,\boldsymbol{e}_y + \frac{\partial V}{\partial z}\,\boldsymbol{e}_z,$$

then (1.77) can be written in vector form as:

$$\boldsymbol{F} = -\operatorname{grad} V. \tag{1.78}$$

If we take the derivative of the first expression in (1.77) with respect to y and the derivative of the second expression with respect to x, then the right-hand sides of the two equations will be the same. Thus one has that $\partial F_x/\partial y = \partial F_y/\partial x$. Cyclic permutation of the coordinates further shows that, if a potential exists for a force, then

$$\frac{\partial F_x}{\partial y} = \frac{\partial F_y}{\partial x}, \quad \frac{\partial F_y}{\partial z} = \frac{\partial F_z}{\partial y}, \quad \frac{\partial F_z}{\partial x} = \frac{\partial F_x}{\partial z}. \tag{1.79}$$

Using these relations, one can easily check if a force $\boldsymbol{F}(x, y, z)$ can emanate from a potential. If we introduce the definition of the *rotation* of a force $\boldsymbol{F}$

$$\mathrm{rot}\,\boldsymbol{F} = \begin{vmatrix} \boldsymbol{e}_x & \boldsymbol{e}_y & \boldsymbol{e}_z \\ \dfrac{\partial}{\partial x} & \dfrac{\partial}{\partial y} & \dfrac{\partial}{\partial z} \\ F_x & F_y & F_z \end{vmatrix} = \left(\dfrac{\partial F_z}{\partial y} - \dfrac{\partial F_y}{\partial z} \right) \boldsymbol{e}_x$$

$$+ \left(\dfrac{\partial F_x}{\partial z} - \dfrac{\partial F_z}{\partial x} \right) \boldsymbol{e}_y + \left(\dfrac{\partial F_y}{\partial x} - \dfrac{\partial F_x}{\partial y} \right) \boldsymbol{e}_z\,,$$

then the conditions (1.79) can be compactly written in vector form as

$$\mathrm{rot}\,\boldsymbol{F} = \boldsymbol{0} \tag{1.80}$$

(*irrotational force field*).

If a force has a potential, then from (1.75) and (1.76) one has

$$\mathrm{d}U = -\,\mathrm{d}V, \tag{1.81}$$

and it follows that the work done by the force is

$$U = \int_{\text{\textcircled{0}}}^{\text{\textcircled{1}}} \mathrm{d}U = -\int_{\text{\textcircled{0}}}^{\text{\textcircled{1}}} \mathrm{d}V = -(V_1 - V_0)\,.$$

The potential energy itself is dependent upon a reference location; its difference between two points \textcircled{0} and \textcircled{1}, however, is independent of the reference location.

Substituting for U in (1.70) leads to the *Conservation of Energy Law*

$$T_1 - T_0 = V_0 - V_1$$

or

$$T_1 + V_1 = T_0 + V_0 = \mathrm{const}\,. \tag{1.82}$$

When the applied forces possess a potential, then the sum of the potential and the kinetic energy remains a constant along the trajectory of the system.

We wish now to review two classes of potentials which we have already considered in Volume 1.

a) Potential of the gravitational force mg at a distance z from the surface of the earth (*gravitational potential* near the earth's surface) :

$$V = mg\,z \; . \tag{1.83}$$

b) Potential of a linear elastic translational spring force (with spring constant k)   at a displacement x and of a linear elastic rotational spring torque (with spring constant k_T) at a rotation φ:

$$V = \frac{1}{2}\,k\,x^2 \; , \quad V = \frac{1}{2}\,k_T\,\varphi^2 \; . \tag{1.84}$$

In contrast to gravitational forces and spring forces, frictional forces do not possess a potential energy. They are non-conservative; the work they perform is path dependent. With motion, systems subject to non-conservative forces dissipate energy which appears as heat. Thus, one also calls such forces *dissipative*. The Conservation of Energy Law (1.82) is no longer valid with such forces. In this case, one needs to use the work-energy theorem (1.70), if one is interested in the work of the forces.

The use of the conservation of energy or the work-energy theorem often occurs when one wishes to find the velocity as a function of position (or vice-versa).

Example 1.14   A point mass slides down a *rough* inclined plane (coefficient of friction μ) from a point A with zero initial velocity. The plane transitions smoothly (with same tangent) onto a *frictionless* circular track (Fig. 1.37a).

At which height h above the apex B of the circular track must the mass start in order that it remains on the track at B?

Solution   The point mass will remain on the track up to the point

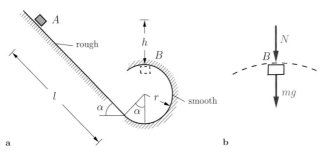

Fig. 1.37

B, if the normal force N only first becomes zero at the apex. From the equation of motion in the radial direction (cf. Fig. 1.37b)

$$\downarrow:\quad m\, a_n = m\,\frac{v_B^2}{r} = mg + N$$

the required velocity at B follows from the condition $N = 0$:

$$v_B^2 = rg\,. \tag{a}$$

The relation between v_B and the desired height h is found from the work-energy theorem. The work of the gravitational forces between A and B is $U_1 = mgh$. The frictional force, which is oriented opposite to the motion, performs work $U_2 = -Rl$. With $R = \mu N = \mu\, mg\cos\alpha$ and $l = (h + r + r\cos\alpha)/\sin\alpha$, it follows that $U_2 = -mg\mu\cot\alpha(h + r + r\cos\alpha)$. The kinetic energy at position A is zero and at B it is $T_B = \frac{1}{2}mv_B^2$. Substituting into (1.70) gives

$$\frac{1}{2}\,mv_B^2 = mgh - mg\mu\cot\alpha\,(h + r + r\cos\alpha)\,.$$

From (a) we can find the desired height:

$$\frac{1}{2}\,mrg = mgh\,(1 - \mu\cot\alpha) - mg\mu r\cot\alpha\,(1 + \cos\alpha)$$

$$\rightarrow\quad h = r\,\frac{\frac{1}{2} + \mu\,(1 + \cos\alpha)\cot\alpha}{1 - \mu\cot\alpha}\,. \tag{b}$$

In the case that the plane is frictionless ($\mu = 0$), we can use the

Conservation of Energy Law (1.82). Noting

$$T_A = 0, \quad T_B = \frac{1}{2} m v_B^2, \quad V_A = mgh, \quad V_B = 0$$

it follows that $\frac{1}{2} m v_B^2 = mgh$ and with relation (a) that

$$\frac{1}{2} mrg = mgh \quad \rightarrow \quad \underline{\underline{h = \frac{1}{2} r}} .$$

The same result also follows from (b), if one simply sets the coefficient of friction to zero.

Example 1.15 A mass m is positioned at a height h above the reference position of a spring (spring constant k); see Fig. 1.38. The mass is released towards the spring with an initial vertical velocity v_0 in a frictionless guide.

Find the maximum compression of the spring.

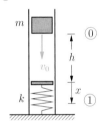

Fig. 1.38

Solution Since both gravitational and elastic spring potentials exist, we can use the Conservation of Energy Law (1.82) to solve this problem. In the initial position ⓪ the mass has kinetic energy $T_0 = m v_0^2 / 2$ and potential energy $V_0 = mgh$ (the reference position is taken at the end of the uncompressed spring). At maximal compression, $x_{\max}$, the mass will be in position ① with kinetic energy $T_1 = 0$. The potential energy at this position will be composed of the potential energy of the spring $\frac{1}{2} k x_{\max}^2$ and the gravitational potential $-mg \, x_{\max}$:

$$V_1 = \frac{1}{2} k x_{\max}^2 - mg \, x_{\max} .$$

Thus conservation of energy says

$$\frac{1}{2}\,mv_0^2 + mgh = 0 + \frac{1}{2}\,k\,x_{\max}^2 - mg\,x_{\max}\,.$$

Solving this quadratic equation gives

$$x_{\max} = \frac{mg}{k}\left[1\,(\pm)\,\sqrt{1 + \frac{kv_0^2}{mg^2} + \frac{2\,hk}{mg}}\,\right].$$

In the special case where $h = 0$ and $v_0 = 0$, it follows that $x_{\max} = 2\,mg/k$. Thus if one suddenly releases a mass directly above an uncompressed spring, then the maximal compression is twice as large as it is in the static case $x_{\text{stat}} = mg/k$ (which can be achieved by slowly releasing the mass).

1.2.8 Universal Law of Gravitation, Planetary and Satellite Motion

In addition to his three fundamental laws (cf. Section 1.2.1), Newton also formulated the *Gravitational Law*. According to this law, between any two masses m and M there is a force (Fig. 1.39a):

$$F = G\frac{Mm}{r^2}\,. \tag{1.85}$$

Here, G is the universal gravitational constant

$$G = 6.673 \cdot 10^{-11}\,\frac{\text{m}^3}{\text{kg\,s}^2}$$

and r is the distance between the masses.

One can show that Newton's gravitational force emanates from a potential. It follows from (1.81) that

$$V = -\int(-F)\,\mathrm{d}r = -G\frac{Mm}{r} + C \tag{1.86}$$

(F is oriented *opposite* $\mathrm{d}r$). If one sets the potential to zero at infinity ($r \to \infty$), then $C = 0$ and

$$V = -G\frac{Mm}{r}\,. \tag{1.87}$$

In the special case that the mass M is the earth, then a mass m above the earth's surface experiences a gravitational force (weight) $F = mg$. If the earth's radius is R, it follows from (1.85) that

$$mg = G\frac{Mm}{R^2} \quad \rightarrow \quad g = \frac{GM}{R^2}\,.$$

Eliminating G in the Gravitational Law, we obtain the gravitational force as a function of the distance from the center of the earth:

$$F = mg\left(\frac{R}{r}\right)^2\,. \tag{1.88}$$

For the potential it follows from (1.86) that

$$V = -\,mg\,\frac{R^2}{r} + C\,.$$

Setting the potential to zero at the earth's surface $r = R$ gives $C = mg\,R$ and with $r = R + z$ (Fig. 1.39b) it follows that

$$V = -\,mg\,\frac{R^2}{R+z} + mg\,R = \frac{mg}{R+z}Rz\,. \tag{1.89}$$

Near the earth's surface ($z \ll R$) the gravitational potential simplifies to (cf. (1.83)):

$$V = mgz\,.$$

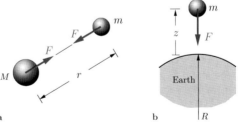

Fig. 1.39   a   b

With Newton's Gravitational Law we can also describe the motion of planets and satellites. Such objects can be treated as point masses as their dimensions are small in comparison to their paths. We will denote the mass of a planet (or satellite) as m and the mass of the sun (respectively the earth) as M. The mass M will be assumed stationary. As the motion of m is typically in a plane, it will be useful to utilize polar coordinates. In this context, Newton's 2nd Law (1.38), along with (1.22) and (1.85), gives in the radial direction

$$m\left(\ddot{r} - r\dot{\varphi}^2\right) = -G\frac{mM}{r^2} \tag{1.90}$$

and in the angular direction

$$m\left(r\ddot{\varphi} + 2\dot{r}\dot{\varphi}\right) = 0 \quad \rightarrow \quad m\frac{1}{r}\frac{d}{dt}(r^2\dot{\varphi}) = 0\,.$$

The second relation expresses *Kepler's 2nd Law* (cf. (1.29a)), whereby the rate of change of the area swept out by the position vector is a constant:

$$r^2\dot{\varphi} = C\,. \tag{1.91}$$

To solve the first equation, we introduce a new variable $u = 1/r$. With (1.91) and $\dot{r} = (dr/d\varphi)\dot{\varphi}$, one has

$$\dot{\varphi} = \frac{C}{r^2} = Cu^2, \quad \dot{r} = \frac{dr}{d\varphi}\frac{C}{r^2} = -C\frac{d}{d\varphi}\left(\frac{1}{r}\right) = -C\frac{du}{d\varphi}\,.$$

Differentiating once more gives

$$\ddot{r} = -C\frac{d^2u}{d\varphi^2}\dot{\varphi} = -C^2u^2\frac{d^2u}{d\varphi^2}\,.$$

Substituting into (1.90) leads to

$$-C^2u^2\frac{d^2u}{d\varphi^2} - \frac{1}{u}C^2u^4 = -GMu^2$$

or after rearrangement

$$\frac{d^2u}{d\varphi^2} + u = \frac{GM}{C^2}\,.$$

This second order inhomogeneous differential equation has the general solution (cf. Chapter 5)

$$u = B \cos (\varphi - \alpha) + \frac{GM}{C^2} .$$

The distance r must therefore satisfy the following relation:

$$r = \frac{1}{u} = \frac{1}{B \cos (\varphi - \alpha) + \dfrac{GM}{C^2}} .$$

Here, B and α are constants of integration. If one measures φ from the point on the trajectory where $\dot{r}$ vanishes, then $\alpha = 0$ and one obtains the path equation

$$r = \frac{p}{1 + \varepsilon \cos \varphi} \tag{1.92a}$$

with

$$p = \frac{C^2}{GM}, \qquad \varepsilon = \frac{BC^2}{GM} . \tag{1.92b}$$

Equation (1.92a) is the focal point relation for a conic section whose type depends on the eccentricity ε. For $\varepsilon < 1$, the path will be an ellipse. This is *Kepler's 1st Law*: Planetary motion is elliptical and the sun is located at a focal point with distance e from the center of the ellipse (Fig. 1.40).

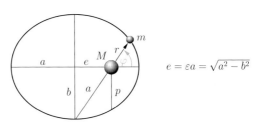

$$e = \varepsilon a = \sqrt{a^2 - b^2}$$

Fig. 1.40

With the constant in the Law of Equal Areas (1.91) and the area of an ellipse $A = \pi\, ab$ (a, b = semi-axes), one can determine the orbital period T:

$$\frac{\mathrm{d}A}{\mathrm{d}t} = \frac{1}{2}r^2\dot{\varphi} = \frac{C}{2} \quad \rightarrow \quad A = \frac{C}{2}\,T \quad \rightarrow \quad T = \frac{2\,A}{C} = \frac{2\,\pi\,ab}{C} .$$

Considering (1.92b), the equation of motion (1.90), and the ellipse parameter $p = b^2/a$ one finds

$$|a_r| = |\ddot{r} - r\dot{\varphi}^2| = \frac{GM}{r^2} = \frac{C^2}{pr^2} = \frac{4\pi^2 a^2 b^2}{T^2 \dfrac{b^2}{a} r^2} = \frac{4\pi^2 a^3}{r^2 T^2}$$

and thus

$$T^2 = \frac{(2\pi)^2 a^3}{GM}. \tag{1.93}$$

This is *Kepler's 3rd Law*: the square of the orbital period T of a planet is proportional to the cube of the semimajor axis of the planet's trajectory.

According to (1.92a), a body in a gravitational field moves on a parabolic path if $\varepsilon = 1$ and on a hyperbolic path if $\varepsilon > 1$. In the computation of satellite motions, one must take into account the gravitational fields due to multiple heavenly bodies (the many body problem).

E1.16

Example 1.16 What is the minimal amount of energy needed to launch a satellite of mass m into a circular orbit with altitude h?

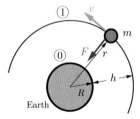

Fig. 1.41

Solution According to Newton's Gravitational Law (1.88), a force $F = mg\,(R^2/r^2)$ acts on the satellite (Fig. 1.41). When the satellite moves in a circular path with distance $r = R + h$ from the earth's center, it has a velocity v that is found from the equation of motion in the radial direction:

$$m\,a_n = F \quad \rightarrow \quad m\frac{v^2}{r} = mg\frac{R^2}{r^2} \quad \rightarrow \quad v^2 = g\frac{R^2}{R+h}.$$

The potential energy of the satellite is found from (1.89). At the surface of the earth and in orbit is takes on the values

$$V_0 = 0 \quad \text{and} \quad V_1 = mg\,\frac{R}{R+h}\,h\,,$$

respectively. The corresponding kinetic energies are

$$T_0 = 0 \quad \text{and} \quad T_1 = \frac{mv^2}{2} = mg\,\frac{R^2}{2\,(R+h)}\,.$$

Thus the minimal amount of energy needed at launch, ΔE, is

$$\underline{\underline{\Delta E}} = E_1 - E_0 = (T_1 + V_1) - (T_0 + V_0)$$

$$= mg\,R\left(\frac{R}{2\,(R+h)} + \frac{h}{R+h}\right) = \underline{\underline{\frac{mg\,R}{2}\left(\frac{R+2\,h}{R+h}\right)}}\,.$$

1.3 Supplementary Examples

Detailed solutions to the following examples are given in (**A**) D. Gross et al. *Formeln und Aufgaben zur Technischen Mechanik 3*, Springer, Berlin 2010, or (**B**) W. Hauger et al. *Aufgaben zur Technischen Mechanik 1-3*, Springer, Berlin 2008.

Example 1.17   A point P moves on a given path from A to B (Fig. 1.42). Its velocity v decreases linearly with the arc-length s from the value v_0 at A to zero at B.
   How long does it take P to reach point B?

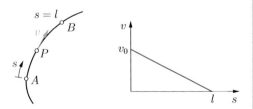

Fig. 1.42

Result: see (**B**) $t_B \to \infty$.

E1.18

Example 1.18 A radar screen tracks a rocket which rises vertically with a constant acceleration a (Fig. 1.43). The rocket is launched at $t = 0$.

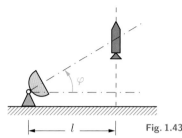

Determine the angular velocity $\dot{\varphi}$ and the angular acceleration $\ddot{\varphi}$ of the radar screen. Calculate the maximum angular velocity $\dot{\varphi}$ and the corresponding angle φ.

Fig. 1.43

Results: see (**B**) $\dot{\varphi}(t) = \dfrac{a\,t}{l} \cdot \dfrac{1}{D}$, $D = \left[1 + \left(\dfrac{a\,t^2}{2l}\right)^2\right]$,

$$\ddot{\varphi}(t) = \left(\frac{a}{l} - \frac{3\,a^3\,t^4}{4\,l^3}\right) \cdot \frac{1}{D^2}, \quad \dot{\varphi}_{\max} = \sqrt{\frac{3\sqrt{3}}{8}\frac{a}{l}}, \quad \varphi = 30°.$$

E1.19

Example 1.19 Two point masses P_1 and P_2 start at point A with zero initial velocities and travel on a circular path. P_1 moves with a uniform tangential acceleration a_{t1} and P_2 moves with a given uniform angular velocity ω_2.

a) What value must a_{t1} have in order for the two masses to meet at point B?

b) What is the angular velocity of P_1 at B?

c) What are the normal accelerations of the two masses at B?

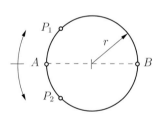

Fig. 1.44

Results: see (**A**) $a_{t1} = \dfrac{2r\omega_2^2}{\pi}$, $\omega_1(t_B) = 2\omega_2$,

$$a_{n1} = 4r\omega_2^2, \quad a_{n2} = r\omega_2^2.$$

E1.20

Example 1.20 A child of mass m jumps up and down on a trampoline in a periodic manner. The child's jumping velocity (upwards) is v_0 and during the contact time Δt the contact force $K(t)$ has a triangular form.

Find the necessary contact force amplitude K_0 and the jumping period T_0.

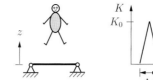

Fig. 1.45

Results: see (**A**) $T_0 = \dfrac{2v_0}{g} + \Delta t$, $K_0 = 2\Big(\dfrac{2v_0}{g\Delta t} + 1\Big)mg$.

Example 1.21 A car is travelling in a circular arc with radius R and velocity v_0 when it starts to brake.

If the tangential deceleration is $a_t(v) = -(a_0 + \kappa v)$, where a_0 and κ are given constants, find the time to brake t_B, the stopping distance s_B, and the normal acceleration a_n during the braking.

E1.21

Fig. 1.46

Results: see (**A**) $t_B = \dfrac{1}{\kappa} \ln\Big(1 + \dfrac{\kappa v_0}{a_0}\Big)$,

$$s_B = \dfrac{a_0}{\kappa^2}\bigg[\dfrac{\kappa v_0}{a_0} - \ln\Big(1 + \dfrac{\kappa v_0}{a_0}\Big)\bigg] ,$$

$$a_n = \dfrac{a_0^2}{R\kappa^2}\bigg\{\Big(1 + \dfrac{\kappa v_0}{a_0}\Big)e^{-\kappa t} - 1\bigg\}^2 .$$

Example 1.22 A point P moves on the quadratic parabola $y = b(x/a)^2$ from A to B. Its position as a function of the time is given by the angle $\varphi(t) = \arctan \omega_0\, t$ (see Fig. 1.47).

Determine the velocity $v(t)$ of point P. How much time elapses until P reaches point B? Calculate its velocity at B.

E1.22

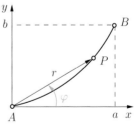

Fig. 1.47

Results: see (**B**)

$$v(t) = \dfrac{a^2}{b}\omega_0\sqrt{1 + 4\omega_0^2\, t^2} , \quad t_B = \dfrac{b}{a\,\omega_0} , \quad v_B = \dfrac{a^2}{b}\omega_0\sqrt{1 + 4\dfrac{b^2}{a^2}} .$$

E1.23

Example 1.23 A rod with length l rotates about support A with angular position given by $\varphi(t) = \kappa t^2$.
A body G slides along the rod with position $r(t) = l(1 - \kappa t^2)$.
a) Find the velocity and acceleration of G when $\varphi = 45°$.
b) At what angle φ does G hit the support?
Given: $l = 2$ m, $\kappa = 0.2$ s^{-2}.

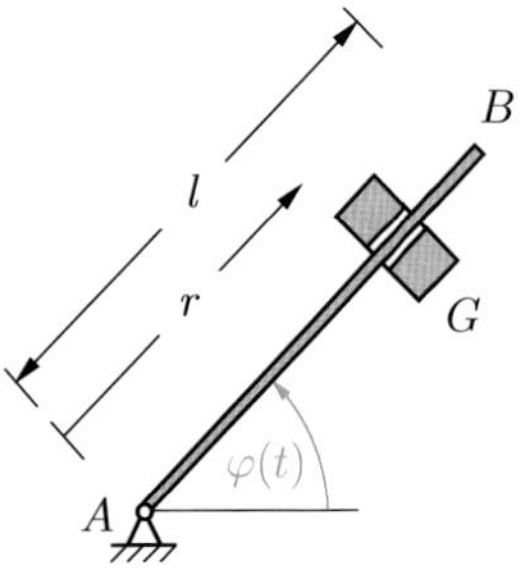

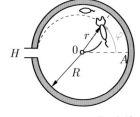

Results: see (**A**)

Fig. 1.48

$$v = 1.62 \, \text{m/s}, \quad a = 2.57 \, \text{m/s}^2, \quad \varphi_E = 1 = (\hat{=}57.3°).$$

E1.24

Example 1.24 A mouse sits in a tower (with radius R) at point A and a cat sits at the center 0.

If the mouse runs at a constant velocity v_M along the tower wall and the cat chases it in an Archimedian spiral $r(\varphi) = R\varphi/\pi$, what must the cat's constant velocity v_C be in order to catch the mouse just as the mouse reaches its escape hole H? At what time does it catch the mouse?

Fig. 1.49

Results: see (**A**) $T = \dfrac{\pi R}{v_M}, \quad v_C = 0.62 \, v_M$.

E1.25

Example 1.25 A soccer player kicks the ball (mass m) so that it leaves the ground at an angle α_0 with the initial velocity v_0 (Fig. 1.50). The air exerts the drag force $F_d = k v$ on the ball; it acts in the direction opposite to the velocity.

Determine the velocity $v(t)$ of the ball. Calculate the horizontal component v_H of v when the ball reaches the team mate at the distance l.

Results: see (**B**) $\dot{x}(t) = v_0 \cos \alpha_0 \, e^{-kt/m}$,

$$\dot{z}(t) = \frac{mg}{k} - \left(\frac{mg}{k} + v_0 \sin \alpha_0\right) e^{-kt/m} , \quad v_H = v_0 \cos \alpha_0 - \frac{k \, l}{m} .$$

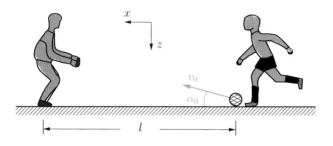

Fig. 1.50

Example 1.26 A body with mass m starts at a height h at time $t = 0$ with an initial horizontal velocity v_0.

If the wind resistance can be approximated by a *horizontal* force $H = c_0\, m\, \dot{x}^2$, at what time t_B and location x_B does it hit the ground?

E1.26

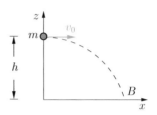

Fig. 1.51

Results: see (**A**) $t_B = \sqrt{\dfrac{2h}{g}}, \quad x_B = \dfrac{1}{c_0} \ln\left(1 + c_0 v_0 \sqrt{\dfrac{2h}{g}}\right).$

Example 1.27 A mass m slides on a rotating frictionless and massless rod S such that it is pressed against a rough circular wall (with coefficient of friction μ).

If the mass starts in contact with the wall at a velocity v_0, how many rotations will it take for its velocity to drop to $v_0/10$?

Result: see (**A**) $n = \dfrac{\ln 10}{2\pi\mu}.$

E1.27

Fig. 1.52

Example 1.28 A car (mass m) is travelling with the constant velocity v along a banked circular curve (radius r, angle of slope α), see Fig. 1.53. The coefficient of static friction μ_0 between the tyres of the car and the surface of the road is given.

E1.28

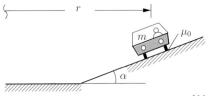

Abb. 1.53

Determine the region of the allowable velocity so that sliding (down or up the slope) does not take place.

Result: see (**B**) $\dfrac{\tan\alpha - \mu_0}{1 + \mu_0\tan\alpha} \leq \dfrac{v^2}{g\,r} \leq \dfrac{\tan\alpha + \mu_0}{1 - \mu_0\tan\alpha}$.

E1.29

Example 1.29 A car (mass m) has the velocity v_0 at the beginning of a curve (Fig. 1.54). Then it slows down with the constant tangential acceleration $a_t = -a_0$. The coefficient of static friction between the road and the tyres is μ_0.

Calculate the velocity v of the car as a function of the arc-length s. What is the necessary radius of curvature $\rho(s)$ of the road so that the car does not slide?

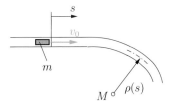

Fig. 1.54

Results: see (**B**) $v(s) = \sqrt{v_0^2 - 2a_0 s}$, $\rho(s) \geq \dfrac{v_0^2 - 2a_0\,s}{\sqrt{\mu_0^2 g^2 - a_0^2}}$.

E1.30

Example 1.30 A bowling ball (mass m) moves with the constant velocity v_0 on the frictionless return of a bowling alley. It is lifted on a circular path (radius r) to the height $2r$ at the end of the return. The upper part of the circular path has a frictionless guide of length $r\,\varphi_G$ (Fig. 1.55).

Given the angle φ_G, determine the velocity v_0 such that the bowling ball reaches the upper level.

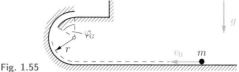

Fig. 1.55

Result: see (**B**)

$$v_0^2 = (2 + 3\cos\varphi_G)\, g\, r \quad \text{for} \quad \varphi_G < \varphi_G^* = \arccos 2/3\,,$$
$$v_0^2 = 4\, g\, r \qquad\qquad \text{for} \quad \varphi_G > \varphi_G^*\,.$$

Example 1.31 A point mass m is subject to a central force $F = mk^2r$, where k is a constant and r is the distance of the mass from the origin 0. At time $t = 0$ the mass is located at P_0 and has velocity components $v_x = v_0$ and $v_y = 0$.

E1.31

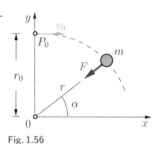

Find the trajectory of the mass.

Fig. 1.56

Result: see (**A**) $\left(\dfrac{x}{v_0/k}\right)^2 + \left(\dfrac{y}{r_0}\right)^2 = 1\,.$

Example 1.32 A centrifuge with radius r rotates with constant angular velocity ω_0. A mass m is to be placed at rest in the centrifuge and accelerated within a time t_1 to the angular velocity ω_0.

E1.32

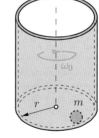

What will be the needed (constant) moment M acting on the mass? What is the power P of this moment?

Results: see (**A**) $M = \dfrac{r^2 m \omega_0}{t_1}\,,\ P = \dfrac{r^2 m \omega_0^2}{t_1}\,.$ **Fig. 1.57**

Example 1.33 A skier (mass m) has the velocity $v_A = v_0$ at point A of the cross country course (Fig. 1.58). Although he tries hard not to lose velocity skiing uphill, he reaches point B with only the velocity $v_B = 2\,v_0/5$. Skiing downhill between point B and

E1.33

the finish C he again gains speed and reaches C with $v_C = 4\,v_0$. Between B and C assume that a constant friction force acts due to the soft snow in this region; the drag force from the air on the skier can be neglected.

Calculate the work done by the skier on the path from A to B (here the friction force is negligible). Determine the coefficient of kinetic friction between B and C.

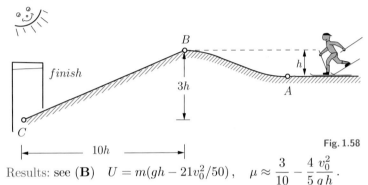

Fig. 1.58

Results: see (**B**) $U = m(gh - 21v_0^2/50)$, $\mu \approx \dfrac{3}{10} - \dfrac{4}{5}\dfrac{v_0^2}{g\,h}$.

E1.34

Example 1.34 A circular disk (radius R) rotates with the constant angular velocity Ω. A point P moves along a straight guide; its distance from the center of the disk is given by $\xi = R\sin\omega\,t$ where $\omega = $ const (Fig. 1.59).

Determine the velocity and the acceleration of P.

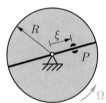

Fig. 1.59

Results: see (**B**) $\boldsymbol{v} = R\omega\cos\omega t\,\boldsymbol{e}_r + R\Omega\sin\omega t\,\boldsymbol{e}_\varphi$,

$$\boldsymbol{a} = -R\,(\omega^2 + \Omega^2)\sin\omega t\,\boldsymbol{e}_r + 2\,R\omega\,\Omega\cos\omega t\,\boldsymbol{e}_\varphi.$$

E1.35

Example 1.35 A chain with length l and mass m hangs over the edge of a frictionless table by an amount e.

If the chain starts with zero initial velocity, find the position of the end of the chain as a function of time.

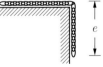

Result: see (**A**) $x(t) = e\cosh\sqrt{\dfrac{g}{l}}\,t$.

Fig. 1.60

Example 1.36 A rotating source of light throws a beam of light onto a screen (Fig. 1.61). Point P on the screen should move with constant velocity v_0.

Determine the required angular acceleration $\ddot{\varphi}(\varphi)$ of the source of light. Display $\dot{\varphi}(\varphi)$ and $\ddot{\varphi}(\varphi)$ in diagrams.

E1.36

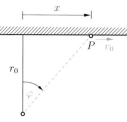

Fig. 1.61

Results: see (**A**) $\dot{\varphi} = \frac{v_0}{r_0} \cos^2 \varphi$, $\ddot{\varphi} = -2\left(\frac{v_0}{r_0}\right)^2 \sin \varphi \cos^3 \varphi$.

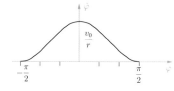

Example 1.37 A car travels with the velocity $v_0 = 100$ km/h. At time $t = 0$, the driver fully applies the brakes. Then the car is sliding on the rough road (coefficient of kinetic friction μ).

Use the simplest model of the car (point mass) and calculate the time t^* and the distance x^* until the car comes to a stop

E1.37

a) on a dry road ($\mu = 0.8$),

b) on a wet road ($\mu = 0.35$).

Results: see (**A**) a) $t^* = 3.55$ s, $x^* = 49$ m;

b) $t^* = 8.1$ s, $x^* = 112$ m.

Example 1.38 In order to determine the coefficient of restitution e experimentally, a ball is dropped from height h_0 onto a horizontal rigid surface (Fig. 1.62). After the ball has hit the surface 7 times, it reaches only 20% of the original height h_0.

Calculate the coefficient of restitution.

E1.38

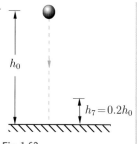

$h_7 = 0.2h_0$

Result: see (**A**) $e = 0.89$.

Fig. 1.62

1.4 Summary

- Velocity is the time derivative of the position vector: $\boldsymbol{v} = \dot{\boldsymbol{r}}$. It is tangential to the trajectory.
- Acceleration is the time derivative of the velocity vector: $\boldsymbol{a} = \dot{\boldsymbol{v}}$.
- For circular motion, the velocity, tangential acceleration, and normal acceleration are given by

$$v = r\,\dot{\varphi}\,, \quad a_t = r\,\ddot{\varphi}\,, \quad a_n = r\,\dot{\varphi}^2 = v^2/r\,.$$

- Newton's 2nd Law: $m\,\boldsymbol{a} = \boldsymbol{F}$.
- In general the following steps are required to determine the motion of a point mass:
  ◇ Sketch a free-body diagram of the point mass.
  ◇ Choose an appropriate coordinate system.
  ◇ Determine the equations of motion.
  ◇ Integrate the equations of motion and use the initial conditions.

- Impulse Law: $m\,\boldsymbol{v} - m\,\boldsymbol{v}_0 = \int\limits_{t_0}^{t} \boldsymbol{F}(\bar{t})\mathrm{d}\bar{t} = \hat{\boldsymbol{F}}$,

  $\boldsymbol{p} = m\,\boldsymbol{v}$ linear momentum.

- Angular momentum theorem: $\dot{\boldsymbol{L}}^{(0)} = \boldsymbol{M}^{(0)}$,

  $\boldsymbol{L}^{(0)} = \boldsymbol{r} \times \boldsymbol{p} = \boldsymbol{r} \times m\,\boldsymbol{v}$ angular momentum with respect to 0,

  $\boldsymbol{r}$ position vector from 0 to the point mass.

- Work-energy theorem: $T_1 - T_0 = U$,

  U work of the forces between trajectory points ⓪ und ①,

  $T = m\,v^2/2$ kinetic energy.

- Conservation of Energy Law: $T + V = $const,

  V potential energy (e.g. mgz, $kx^2/2$, $k_T\varphi^2/2$).

  N.B.: all forces must possess a potential (conservative system).

Chapter 2

Dynamics of Systems of Point Masses

2

2 Dynamics of Systems of Point Masses

———— Objectives: Up to now we have concerned ourselves with the study of single point masses. We now wish to extend the ideas developed in Chapter 1 to systems of interacting point masses. This includes the concepts of *linear* and *angular momentum*, the *impulse law*, and the *work-energy theorem*. The reader will learn how one studies such systems and how to systematically apply the laws of motion to them.

2.1 Fundamentals

Having concentrated so far on point masses, we now wish to study systems of point masses. A *system of point masses* is understood to be a finite collection of point masses that interact with each other.

In natural and engineered systems one often finds systems composed of *multiple* bodies that can be idealized as point masses for the purpose of the analysis of their motion. In other problems, one can idealize *single* bodies as being composed of multiple point masses. From this point of view, one also recognizes that an understanding of the behavior of a system of point masses is a stepping stone on the way to understanding the behavior of continuum bodies with distributed mass.

In discussing the interactions of point masses in a system, one distinguishes between two classes of interactions: kinematic couplings and physical couplings. Typical kinematic couplings are of the form of *kinematic constraints*, which are relations between the coordinates of the masses. They are given by so-called *geometric* or *kinematic constraint equations*. A simple example is shown in Fig. 2.1a, where the two masses are connected by an inextensible, massless rope. Let x_1 and x_2 be their vertical displacements from arbitrary reference positions, then the vertical motion of one completely specifies the motion of the other (excluding the possibility of horizontal motions). In this example the geometric constraint equation is simply $x_1 = x_2$.

If the distance between two points in a system does not change, then the constraint is termed a *rigid constraint*. As a simple example, consider the dumbbell shown in Fig. 2.1b. The two point masses m_1 and m_2 are connected by a rigid, massless rod which enforces a constant separation l between them. This relation can be expressed by the geometric relation (kinematic constraint):

$$(x_2 - x_1)^2 + (y_2 - y_1)^2 + (z_2 - z_1)^2 = l^2 \,. \tag{2.1}$$

From the number of masses and the number of kinematic constraints one can determine the number of degrees of freedom, f, of a system. This latter number, tells us how many independent

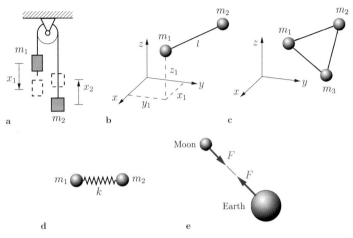

Fig. 2.1

coordinates are needed to uniquely specify the configuration of a system – that is the location of each mass in the system. In the example shown in Fig. 2.1a, there is only one degree of freedom. Of the two coordinates x_1 and x_2, only one can be independently specified due to the geometric constraint equation $x_1 = x_2$. For the system shown in Fig. 2.1b, there are five degrees of freedom. The locations of the two masses are given by $2 \cdot 3 = 6$ coordinates (three for each point mass in three-dimensional space); however, because of the geometric constraint equation (2.1), one only needs to specify five of them ($f = 2 \cdot 3 - 1 = 5$ degrees of freedom) in order to determine all six. This could include the three independent translations (in the x-, y-, and z-directions) and rotations about two axes (neither co-linear with the dumbbell axis). Note that a rotation about the dumbbell axis does not produce a change in the configuration of the point masses and is thus not a degree of freedom of the system.

In general, the number of degrees of freedom, f, of a system composed of n masses in three-dimensional space is given by the number of individual mass coordinates $3n$ minus r, the number of kinematic constraints:

$$f = 3\,n - r\,. \tag{2.2}$$

Accordingly, the 3-mass system shown in Fig. 2.1c with three rigid constraints has $f = 3 \cdot 3 - 3 = 6$ degrees of freedom. If additional point masses are rigidly connected to this system, the number of degrees of freedom will remain at six as the rigid couplings will not permit additional degrees of freedom. As a consequence, a rigid body, which can be thought of as being composed of an infinite number of point masses, has six degrees of freedom in three-dimensional space.

For planar motions, i.e. those restricted to a two-dimensional space, the number of degrees of freedom is given by

$$f = 2\,n - r\,. \tag{2.3}$$

For a 3-mass system with three rigid constraints in a plane, there are $f = 2 \cdot 3 - 3 = 3$ degrees of freedom. Analogous to the discussion above, a rigid body constrained to move in a two-dimensional space, will have three degrees of freedom.

In contrast to a kinematical coupling, a *physical coupling* defines a relation between the positions of the masses of a system and the forces acting between them; see, for example, the spring coupling in Fig. 2.1d and the gravitational coupling in Fig. 2.1e. In these examples, the physical coupling is given by force-separation relations – the spring law (Hooke's law) and the gravitational law (Newton's law of gravitation), respectively.

In what follows, we will consider a system of n point masses $m_i\,(i = 1, ..., n)$ in three-dimensional space with arbitrary couplings (Fig. 2.2). Conceptually, these masses will be separated from masses outside of the system by an imaginary *system boundary* that encloses all n of them.

Each point mass m_i is subject to both internal as well as external forces. The *external forces* $\boldsymbol{F}_i$ emanate from outside the system boundary. They can be applied forces (e.g. weight) or reaction forces (e.g. support or constraint forces). The index i indicates

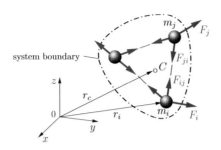

Fig. 2.2

that force $\boldsymbol{F}_i$ acts on mass m_i. The *internal forces* $\boldsymbol{F}_{ij}$ act between the masses of the system. The indices on $\boldsymbol{F}_{ij}$ indicate a force that mass m_j exerts on mass m_i. Alternately, $\boldsymbol{F}_{ji}$ is a force that m_i exerts on m_j. The line of action of the internal forces is always directed along the connecting line between the two masses in question. Because "actio = reactio" (Newton's 3rd Law), $\boldsymbol{F}_{ij}$ and $\boldsymbol{F}_{ji}$ must have the same magnitude and be oppositely directed:

$$\boldsymbol{F}_{ji} = -\boldsymbol{F}_{ij} \, . \tag{2.4}$$

The motion of the masses of a system can be determined by applying Newton's 2nd Law (1.38) to each mass m_i. With position vectors $\boldsymbol{r}_i$, it follows that

$$m_i \ddot{\boldsymbol{r}}_i = \boldsymbol{F}_i + \sum_j \boldsymbol{F}_{ij} \, , \quad (i = 1, \ldots, n) \, . \tag{2.5}$$

The sum over j includes all internal forces acting on m_i. Additionally, one needs to employ the kinematic and physical coupling relations that express interactions between the masses for a complete system of equations.

E2.1

Example 2.1 The system shown in Fig. 2.3a consists of two weights $W_1 = m_1 g$ and $W_2 = m_2 g$ that are connected by an inextensible, massless rope via a set of massless pulleys.

Find the accelerations of the masses and the forces in the rope when the system is released from rest.

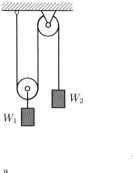 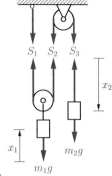

Fig. 2.3

a

b

Solution The system is first separated using a free-body diagram and the internal forces, S_i, and the external forces, W_i, are explicitly drawn-in (Fig. 2.3b). If we assume that W_2 moves downwards, then W_1 will move upwards. Thus we will measure the positions of the masses using coordinates, x_1 and x_2, that are taken as positive in opposite directions. In this case, the equations of motion (2.5) for m_1 and m_2 read

$$m_1 \ddot{x}_1 = -m_1 g + S_1 + S_2, \quad m_2 \ddot{x}_2 = m_2 g - S_3 \,.$$

Here, the internal forces in the rope correspond to the coupling forces in the system. Since the pulleys are assumed to be massless, $S_1 = S_2 = S_3 = S$. Thus it follows that

$$m_1 \ddot{x}_1 = -m_1 g + 2S, \quad m_2 \ddot{x}_2 = m_2 g - S \,. \tag{a}$$

The coordinates x_1 and x_2 are not independent: the system has only *a single* degree of freedom. If m_2 moves downwards a distance x_2, then m_1 will move upwards half that distance. The kinematic constraint can be expressed as

$$x_1 = \frac{1}{2} x_2 \quad \rightarrow \quad \dot{x}_1 = \frac{1}{2} \dot{x}_2 \quad \rightarrow \quad \ddot{x}_1 = \frac{1}{2} \ddot{x}_2 \,. \tag{b}$$

Between (a) and (b), we have three equations for the determination of the three unknowns $\ddot{x}_1, \ddot{x}_2$, and S. Solving, gives

$$\ddot{x}_1 = \frac{1}{2}\ddot{x}_2 = g\,\frac{2\,m_2 - m_1}{m_1 + 4\,m_2}, \qquad S = \frac{3\,m_1\,m_2\,g}{m_1 + 4\,m_2}.$$

If $W_1 = 2\,W_2$, then $\ddot{x}_1 = \ddot{x}_2 = 0$, $S = W_2$, and the system is seen to be in equilibrium.

2.2 Linear Momentum for a System of Point Masses

From the equations of motion (2.5) for the *individual* point masses (cf. Fig. 2.2)

$$m_i\,\ddot{r}_i = F_i + \sum_j F_{ij}\,,$$

the laws of motion that hold for the *whole* system can be derived.

If we sum the equations of motion over all n masses, then we have

$$\sum_i m_i\,\ddot{r}_i = \sum_i F_i + \sum_i \sum_j F_{ij}\,. \tag{2.6}$$

The double sum on the right-hand side includes *all internal* forces acting between the masses in the system. As these are pairwise equal and opposite ($F_{ij} = -F_{ji}$), the double sum is zero. Thus,

$$\sum_i m_i\,\ddot{r}_i = F, \tag{2.7}$$

where $F = \sum_i F_i$ is the resultant force of all the *external* forces acting on the system.

In order to simplify the left-hand side of (2.7), we introduce the position vector r_c for the system's *center of mass* or center of gravity C (cf. Fig. 2.2 and Volume 1, Chapter 4):

$$r_c = \frac{1}{m}\sum_i m_i\,r_i \quad \rightarrow \quad m r_c = \sum_i m_i\,r_i\,. \tag{2.8}$$

Here, $m = \sum_i m_i$ is the total mass of the system. Taking two time derivatives of (2.8) and using definitions $\boldsymbol{v} = \dot{\boldsymbol{r}}$ and $\boldsymbol{a} = \dot{\boldsymbol{v}} = \ddot{\boldsymbol{r}}$, yields

$$m\,\boldsymbol{v}_c = \sum_i m_i\,\boldsymbol{v}_i \quad \text{and} \quad m\,\boldsymbol{a}_c = \sum_i m_i\,\ddot{\boldsymbol{r}}_i\,. \tag{2.9}$$

Substituting into (2.7) yields the equation of motion for the center of mass:

$$m\,\boldsymbol{a}_c = \boldsymbol{F}\,. \tag{2.10}$$

It has the same form as the equation of motion (1.38) for a single point mass. In words, (2.10) says:

> The center of mass of a system moves as though it were a point mass (with same total mass) subject to the totality of forces acting on the whole system.

The equation of motion (2.10) is known as the *law of motion for the center of mass*. Note that the internal forces have no influence on the motion of the center of mass.

The vector equation (2.10) corresponds to three scalar equations – one for each component. For example, in Cartesian coordinates

$$m\,\ddot{x}_c = F_x\,, \quad m\,\ddot{y}_c = F_y\,, \quad m\,\ddot{z}_c = F_z\,.$$

The total linear momentum $\boldsymbol{p} = \sum_i \boldsymbol{p}_i = \sum_i m_i\,\boldsymbol{v}_i$ of the system can be expressed using (2.9) as

$$\boldsymbol{p} = m\,\boldsymbol{v}_c\,. \tag{2.11}$$

Thus, the total linear momentum can be determined from the product of the total mass m and the center of mass velocity $\boldsymbol{v}_c$.

If we take the time derivative of (2.11) and substitute into (2.10), then we see that

$$\dot{\boldsymbol{p}} = \boldsymbol{F}. \tag{2.12}$$

In words: the time rate of change of the total linear momentum is equal to the total resultant of the external forces. As with the single point mass case, we can integrate (2.12) with respect to time to give an *Impulse Law*

$$\boldsymbol{p} - \boldsymbol{p}_0 = \int_{t_0}^{t} \boldsymbol{F} \, \mathrm{d}\bar{t} = \hat{\boldsymbol{F}}, \tag{2.13}$$

where we have used the initial condition $\boldsymbol{p}_0 = \boldsymbol{p}(t_0)$. The difference in the linear momentum between two moments in time is equal to the linear impulse – i.e. the time integral of the external forces acting on the system, $\hat{\boldsymbol{F}}$.

In the special case that the external resultant is zero ($\boldsymbol{F} = \boldsymbol{0}$), then (2.13) gives

$$\boldsymbol{p} = m \, \boldsymbol{v}_c = \boldsymbol{p}_0 = \mathrm{const}. \tag{2.14}$$

The linear momentum of the system is a constant (*conservation of linear momentum*) and the center of mass moves uniformly and in a straight line. Relation (2.14) is known as the *principle of conservation of linear momentum*.

Example 2.2 A mass m in zero gravity moves with velocity v at an angle $\alpha = 30°$ with respect to the horizontal. The mass suddenly splits into three equal pieces $m_1 = m_2 = m_3 = m/3$ (Fig. 2.4). After splitting, masses m_1 and m_2 travel at angles $\beta_1 = 60°$ and $\beta_2 = 90°$, respectively. Mass m_3 stays at rest. Determine the velocities v_1 and v_2.

Solution The mass m, before the split, and the masses m_1, m_2, and m_3 afterwards will constitute our system. No external forces are acting on the system, thus according to (2.14) the linear mo-

E2.2

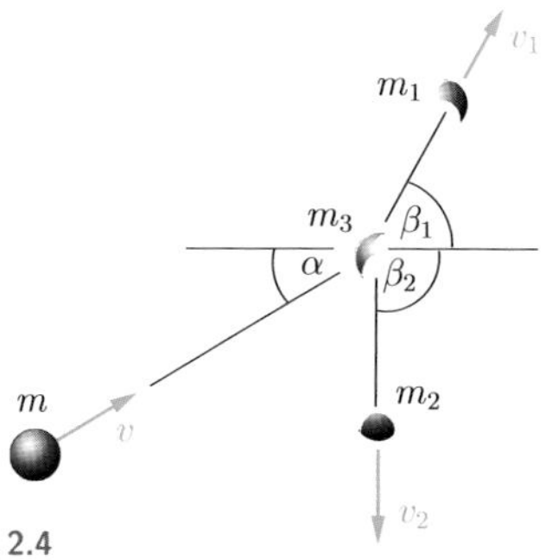

Fig. 2.4

mentum of the system does not change (linear momentum before splitting = linear momentum after splitting). Componentwise, the principle of conservation of linear momentum gives

$$\rightarrow: \quad m\, v \cos\alpha = m_1\, v_1 \cos\beta_1,$$

$$\uparrow: \quad m\, v \sin\alpha = m_1\, v_1 \sin\beta_1 - m_2\, v_2 \sin\beta_2\,.$$

Thus, one has two equations for the two unknowns v_1 and v_2. Solving gives

$$\underline{\underline{v_1}} = v\,\frac{m \cos\alpha}{m_1 \cos\beta_1} = \underline{\underline{3\sqrt{3}\,v}},$$

$$\underline{\underline{v_2}} = \frac{m_1\, v_1 \sin\beta_1 - m\, v \sin\alpha}{m_2 \sin\beta_2} = \underline{\underline{3\,v}}\,.$$

2.3 Angular Momentum Theorem for a System of Point Masses

According to the angular momentum theorem (1.63), for each point mass m_i, one has that $\dot{\boldsymbol{L}}_i^{(0)} = \boldsymbol{M}_i^{(0)}$. Noting that m_i is subject to external forces $\boldsymbol{F}_i$ as well as internal forces $\boldsymbol{F}_{ij}$ (cf.

Fig. 2.2), one finds

$$(\boldsymbol{r}_i \times m_i \boldsymbol{v}_i)^{\cdot} = \boldsymbol{r}_i \times \boldsymbol{F}_i + \sum_j \boldsymbol{r}_i \times \boldsymbol{F}_{ij} \,.$$

Summing over all n masses gives

$$\sum_i (\boldsymbol{r}_i \times m_i \boldsymbol{v}_i)^{\cdot} = \sum_i \boldsymbol{r}_i \times \boldsymbol{F}_i + \sum_i \sum_j \boldsymbol{r}_i \times \boldsymbol{F}_{ij} \,. \qquad (2.15)$$

The left-hand side of this relation is simply the time derivative of the total angular momentum of the system

$$\boldsymbol{L}^{(0)} = \sum_i \boldsymbol{L}_i^{(0)} = \sum_i (\boldsymbol{r}_i \times m_i \boldsymbol{v}_i) \qquad (2.16)$$

with respect to the fixed point 0. Since the internal forces are pairwise equal and opposite, $\boldsymbol{F}_{ij} = -\boldsymbol{F}_{ji}$, and act along the line connecting masses m_i and m_j, the double sum on the right-hand side of (2.15) can be shown to be zero – only the total moment of the external forces remains:

$$\boldsymbol{M}^{(0)} = \sum_i \boldsymbol{M}_i^{(0)} = \sum_i \boldsymbol{r}_i \times \boldsymbol{F}_i \,. \qquad (2.17)$$

Thus we find from (2.15) the *angular momentum theorem* for a system of point masses:

$$\dot{\boldsymbol{L}}^{(0)} = \boldsymbol{M}^{(0)} \,. \qquad (2.18)$$

The theorem states that the time rate of change of the total angular momentum of a system of point masses relative to a *fixed* point 0 is equal to the resultant moment of all the external forces about the same point.

If the resultant external moment is zero ($\boldsymbol{M}^{(0)} = \boldsymbol{0}$), then $\dot{\boldsymbol{L}}^{(0)} = \boldsymbol{0}$ and the angular momentum is constant (*conservation of angular momentum*).

As an important special case, let us examine a system of point masses rotating about a fixed axis a-a to which they are rigidly attached (Fig. 2.5). Without loss of generality, we place the origin

0 on the axis of rotation and align the z-axis with it. Following Section 1.2.6, the z-component of the angular momentum of a mass m_i is given by

$$L_{iz} = L_{ia} = m_i\, r_i^2\, \dot\varphi\,. \tag{2.19}$$

Here, r_i is the orthogonal distance of m_i from the axis of rotation. For the components L_{iz} and L_{ia}, we have replaced the superscript denoting the point of reference by a second subscript denoting the axis of rotation (here z and a-a, respectively). This notational convention can also be applied to the components of moments and will often be used in what follows.

As all the masses move with the same angular velocity $\dot\varphi$, summing (2.19) over all masses gives

$$L_z = L_a = \sum_i L_{ia} = \sum_i m_i\, r_i^2\, \dot\varphi = \Theta_a\, \dot\varphi\,. \tag{2.20}$$

The variable

$$\Theta_a = \sum_i m_i\, r_i^2 \tag{2.21}$$

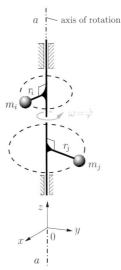

Fig. 2.5

is called the *mass moment of inertia* of the system relative to the rotation axis *a-a*.

If one observes that $\Theta_a = $ const (rigid constraints), then the time derivative of (2.20) gives with the help of (2.18)

$$\Theta_a \ddot{\varphi} = M_a . \tag{2.22}$$

This equation of motion for the rotation of a rigid system of point masses about a fixed axis is analogous to the equation of motion for the translation of a point mass m (e.g. $m\ddot{x} = F_x$). In place of mass, we have the mass moment of inertia, in place of acceleration we have angular acceleration, and in place of force we have moment (cf. Table 3.1).

When applying (2.22), one should pay attention to the assumed positive sense of rotation for φ and moment M_a. If, for example, φ is taken as positive for clockwise rotation, then M_a should also be measured as positive for clockwise moments about *a-a* and vice-versa.

E2.3

Example 2.3 A pendulum consisting of a rigid, massless rod with two masses m_1 and m_2 is suspended from a frictionless pivot A (Fig. 2.6a). If the system is displaced from the equilibrium position and released, then it will oscillate under the action of gravity in the indicated plane.

Determine the equations of motion for the pendulum.

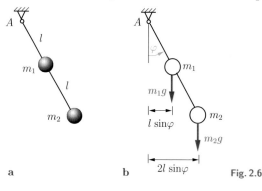

a   b   Fig. 2.6

Solution   The system executes a pure rotation about an axis a through the *fixed* point A. To describe its motion we will use the angular momentum theorem. The angle φ will be taken as positive for counter-clockwise rotations with the reference position being the equilibrium position (the vertical position); see Fig. 2.6b.

With the mass moment of inertia

$$\Theta_a = m_1 \, l^2 + m_2(2\,l)^2 = (m_1 + 4\,m_2)\,l^2$$

and the moment of the external forces (here the weights) about the axis a (mind the positive sense of rotation)

$$M_a = -\,m_1 \, gl \sin\varphi - m_2 \, g(2\,l \sin\varphi) = -\,lg(m_1 + 2\,m_2)\sin\varphi \,,$$

one obtains the equation of motion from (2.22) as

$$(m_1 + 4\,m_2)l^2\ddot{\varphi} = -\,lg(m_1 + 2\,m_2)\sin\varphi$$

$$\rightarrow \quad \underline{\underline{\ddot{\varphi} + \frac{g}{l}\,\frac{m_1 + 2\,m_2}{m_1 + 4\,m_2}\sin\varphi = 0}}\,.$$

For small angles ($\sin\varphi \approx \varphi$), this equation describes a harmonic oscillation (cf. Chapter 5).

2.4 Work-Energy Theorem and Conservation of Energy for a System of Point Masses

Following Section 1.2.7, the work-energy theorem for a single mass m_i in a system of point masses says

$$T_{1i} - T_{0i} = U_i \,, \tag{2.23}$$

where $T_{1i} = m_i v_i^2/2$ is the kinetic energy of mass m_i at a time t_1 and T_{0i} is the kinetic energy at an initial time t_0; U_i is the work of the forces acting on m_i between times t_0 and t_1. With the notation $\boldsymbol{F}_i$ for the external forces and $\boldsymbol{F}_{ij}$ for the internal forces, we can write the work as

$$U_i = \int_{\boldsymbol{r}_{0i}}^{\boldsymbol{r}_{1i}} \left(\boldsymbol{F}_i + \sum_j \boldsymbol{F}_{ij} \right) \cdot \mathrm{d}\boldsymbol{r}_i = U_i^{(e)} + U_i^{(i)}, \tag{2.24}$$

where $U_i^{(e)} = \int \boldsymbol{F}_i \cdot \mathrm{d}\boldsymbol{r}_i$ is the work of the external forces and $U_i^{(i)} = \int \sum \boldsymbol{F}_{ij} \cdot \mathrm{d}\boldsymbol{r}_i$ is the work of the internal forces.

Using the definitions $U = \sum U_i$ and $T = \sum T_i$ and summing (2.23) over all n masses, yields the work-energy theorem for a system of point masses:

$$T_1 - T_0 = U^{(e)} + U^{(i)} = U. \tag{2.25}$$

The sum of the work of all the external and internal forces is equal to the change in the total kinetic energy of the system.

For *rigid* constraints the work of the internal forces, $U^{(i)}$, is zero. In order to show this, consider masses m_i and m_j and the internal

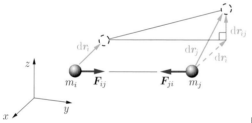

Fig. 2.7

forces acting between them: $\boldsymbol{F}_{ij}$ and $\boldsymbol{F}_{ji} = -\boldsymbol{F}_{ij}$ (Fig. 2.7). For *infinitesimal* displacements $\mathrm{d}\boldsymbol{r}_i$ and $\mathrm{d}\boldsymbol{r}_j$, it follows that

$$\mathrm{d}\boldsymbol{r}_j = \mathrm{d}\boldsymbol{r}_i + \mathrm{d}\boldsymbol{r}_{ij},$$

where $\mathrm{d}\boldsymbol{r}_{ij}$ must be orthogonal to the line connecting the two masses due to their fixed separation; thus $\mathrm{d}\boldsymbol{r}_{ij}$ is also orthogonal to $\boldsymbol{F}_{ji}$. The work $\mathrm{d}U_{ij}^{(i)}$ of both forces is then given as

$$\mathrm{d}U_{ij}^{(i)} = \boldsymbol{F}_{ij} \cdot \mathrm{d}\boldsymbol{r}_i + \boldsymbol{F}_{ji} \cdot \mathrm{d}\boldsymbol{r}_j = \boldsymbol{F}_{ji} \cdot \mathrm{d}\boldsymbol{r}_{ij} = 0.$$

The work $U_{ij}^{(i)} = \int dU_{ij}^{(i)}$ for a finite motion is thus also zero; this result further holds for all other internal forces in the system. For systems composed of only rigid constraints, the work-energy theorem reads

$$T_1 - T_0 = U^{(e)} = U. \tag{2.26}$$

If the external and internal forces are *conservative forces*, i.e. are derivable from potentials $V^{(e)}$ and $V^{(i)}$ (e.g. gravitational forces, spring forces), then the work of the forces is equal to the negative of the difference in the potentials at times t_1 and t_0:

$$U^{(e)} = -(V_1^{(e)} - V_0^{(e)}), \quad U^{(i)} = -(V_1^{(i)} - V_0^{(i)}).$$

Substituting into (2.25) results in the *conservation of energy law*

$$T_1 + V_1^{(e)} + V_1^{(i)} = T_0 + V_0^{(e)} + V_0^{(i)} = \text{const}. \tag{2.27}$$

This equation states that the sum of the kinetic energy and the potential energy is an invariant of the motion. In this situation, one calls the system a *conservative system*. If the internal forces do no work (e.g. rigid constraints), then $U^{(i)} = -(V_1^{(i)} - V_0^{(i)}) = 0$, and from (2.27) we have

$$T_1 + V_1^{(e)} = T_0 + V_0^{(e)} = \text{const}. \tag{2.28}$$

Example 2.4 The system in Fig. 2.8a (cf. Example 2.1) is released from rest.

E2.4

Assuming that the rope is massless and inextensible, and that the pulleys are massless, find the velocity of mass m_1 as a function of its displacement.

Solution Since only conservative forces (weights) are acting on the system and the internal forces do not contribute to the work (inextensible rope), we can employ the conservation of energy relation (2.28). We measure the coordinates x_1 and x_2 (Fig. 2.8b) from the initial position and assume the potentials to be zero for

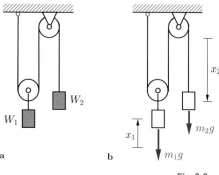

Fig. 2.8

$x_1 = x_2 = 0$. Accounting for the kinematic constraint

$$x_2 = 2 x_1 \quad \rightarrow \quad \dot{x}_2 = 2 \dot{x}_1$$

we have that

$$V_0^{(e)} = 0, \quad V_1^{(e)} = m_1 \, g \, x_1 - m_2 \, g \, x_2 = (m_1 - 2 \, m_2) \, g \, x_1,$$

$$T_0 = 0, \quad T_1 = \frac{1}{2} \, m_1 \, \dot{x}_1^2 + \frac{1}{2} \, m_2 \, \dot{x}_2^2 = \frac{1}{2} \, (m_1 + 4 \, m_2) \, \dot{x}_1^2 \, .$$

Substituting into (2.28) gives the velocity $\dot{x}_1$ as a function of position x_1:

$$\frac{1}{2} \, (m_1 + 4 \, m_2) \, \dot{x}_1^2 + (m_1 - 2 \, m_2) \, g \, x_1 = 0$$

$$\rightarrow \quad \dot{x}_1 = \pm \sqrt{2 \, \frac{2 \, m_2 - m_1}{m_1 + 4 \, m_2} \, g \, x_1} \, .$$

Since the term under the radical must be positive, if $2 \, m_2 > m_1$, then x_1 must also be positive (m_1 moves upwards). In this case the positive sign for the square-root applies. In the case that $2 \, m_2 < m_1$, then x_1 must be negative and the negative sign for the square-root applies ($\dot{x}_1 < 0$).

2.5 Central Impact

The sudden collision of two bodies which causes a change in their velocities is called an *impact*. During a very short period of time large forces are exerted on the bodies. Since these forces lead to time-dependent deformations in the vicinity of the points of contact, a comprehensive treatment of a problem involving impact is rather difficult. However, applying several idealizations will allow us to determine the changes of the velocities in a relatively simple way. We assume:

a) The impact duration t_i is so small that the changes in the positions of the bodies during the impact can be neglected.

b) The impulsive forces at the points of contact are so large that all the other forces (e.g., the weights of the bodies) can be neglected during impact.

c) The deformations of the bodies are so small that they may be neglected when describing the changes of the velocities (i.e., the bodies are assumed to be rigid when the laws of motion are formulated).

Fig. 2.9a shows two bodies during impact. The point of contact P lies in the plane of contact. The normal to the contact plane that passes through P is called the *line of impact*. We refer to a collision as being *direct* if the velocities of the points of contact of both bodies have the direction of the line of impact immediately before the impact. An impact that is not direct is called *oblique*. If the line connecting the centers of mass of the two bodies coincides with the line of impact, the impact is called *central*, otherwise it is *eccentric*. In this section we restrict ourselves to central impact

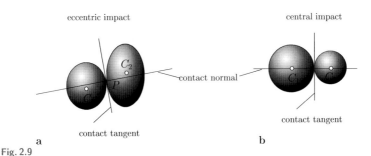

Fig. 2.9

as it occurs, for example, in the collision of two spheres (Fig. 2.9b).

At first let us consider problems involving direct impact. Fig. 2.10a shows two bodies, modelled as point masses m_1 and m_2 which move along a straight line with velocities v_1 and v_2, respectively ($v_1 > v_2$). At time $t = 0$, contact begins. The force $F(t)$ which the two bodies exert on each other first increases with time (Fig. 2.10b). It attains its maximum value at time $t = t^*$. During the time interval $t < t^*$, the *compression phase*, both bodies are compressed in the vicinity of P. At time t^* (largest compression), both masses have the same velocity v^*.

Subsequently, the deformations are partially reduced or completely removed and the impulsive force decreases. The *restitution phase* ends at $t = t_i$ where t_i is the impact duration. Afterwards the force F is zero and the two masses move independently with the velocities $\bar{v}_1$ and $\bar{v}_2$ (Fig. 2.10a).

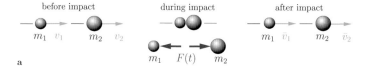

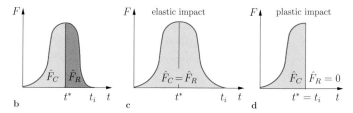

Fig. 2.10

Since the precise time variation of $F(t)$ (Fig. 2.10b) during the impact is usually unknown, we apply the concept of *linear impulse* $\hat{F}$ as we did in Section 1.2.5. We write down the linear impulse for the compression phase and for the restitution phase, respectively:

$$\hat{F}_C = \int_0^{t^*} F(t)\, \mathrm{d}t, \quad \hat{F}_R = \int_{t^*}^{t_i} F(t)\, \mathrm{d}t. \tag{2.29}$$

The total linear impulse is then given by

$$\hat{F} = \int_0^{t_i} F(t)\, \mathrm{d}t = \hat{F}_C + \hat{F}_R. \tag{2.30}$$

If the two bodies behave *ideally elastic*, the linear impulses in the compression phase and in the restitution phase are assumed to be equal: $\hat{F}_C = \hat{F}_R$ (Fig. 2.10c). On the other hand, if the bodies display an *ideally plastic* behavior, the deformations experienced during the compression phase are permanent. The force F is then reduced to zero at $t = t^*$ (Fig. 2.10d) and we obtain $\hat{F}_R = 0$. In this case, both masses have the same velocity v^* after the collision.

A real body responds in a fashion intermediate to the two limiting cases. A *partially elastic* impact is modelled by (cf. (1.55))

$$\hat{F}_R = e\, \hat{F}_C \quad \text{with} \quad 0 \leqq e \leqq 1. \tag{2.31}$$

The constant e is the *coefficient of restitution*. It depends on the materials of the bodies, their form, and to a certain extent on the velocities. It can be determined experimentally. In the case of an *ideal-elastic* impact we have $e = 1$, an *ideal-plastic* impact is characterized by $e = 0$, and for a *partially elastic* impact we have $0 < e < 1$ (see (1.56)). Table 2.1 displays several values of the coefficient of restitution for two spheres made of the same material.

During an impact, the masses experience sudden changes in their velocities (the changes in the positions are negligible). To determine the changes of the velocities we apply the Impulse Law to both masses. Note that the forces and therefore also the linear impulses which are exerted on m_1 and m_2 are of the same magnitude but of opposite directions (action = reaction). Thus,

Table 2.1

material	coefficient of restitution e
wood/wood	≈ 0.5
steel/steel	$0.6\ldots0.8$
glass/glass	0.94
cork/cork	$0.5\ldots0.6$

the Impulse Law (1.51) for the compression phase is given by

$$m_1(v^* - v_1) = -\hat{F}_C,$$
$$m_2(v^* - v_2) = +\hat{F}_C \tag{2.32}$$

and the Impulse Law for the restitution phase reads

$$m_1(\bar{v}_1 - v^*) = -\hat{F}_R,$$
$$m_2(\bar{v}_2 - v^*) = +\hat{F}_R. \tag{2.33}$$

If the coefficient of restitution e is known, we have five equations (2.31) - (2.33) for the five unknowns $\bar{v}_1, \bar{v}_2, v^*, \hat{F}_C$ and $\hat{F}_R$. They can be solved, for example, for the velocities *after* the impact:

$$\bar{v}_1 = \frac{m_1\,v_1 + m_2\,v_2 - e\,m_2(v_1 - v_2)}{m_1 + m_2},$$
$$\bar{v}_2 = \frac{m_1\,v_1 + m_2\,v_2 + e\,m_1(v_1 - v_2)}{m_1 + m_2}. \tag{2.34}$$

In the case of an ideal-plastic impact $(e = 0)$, (2.34) yields

$$\bar{v}_1 = \bar{v}_2 = \frac{m_1\,v_1 + m_2\,v_2}{m_1 + m_2}.$$

This is the velocity v^* at the end of the compression phase.

An ideal-elastic impact $(e = 1)$ leads to

$$\bar{v}_1 = \frac{2\,m_2\,v_2 + (m_1 - m_2)v_1}{m_1 + m_2}, \quad \bar{v}_2 = \frac{2\,m_1\,v_1 + (m_2 - m_1)v_2}{m_1 + m_2}\,.$$

In the special case of two equal masses $(m_1 = m_2 = m)$ this results in

$$\bar{v}_1 = v_2, \quad \bar{v}_2 = v_1\,.$$

In this case an exchange of the velocities takes place. If, for example, mass m_2 is at rest before an impact, then it takes on the initial velocity of m_1 after the impact, whereas m_1 will be rest.

Regardless of the type of impact, the total momentum of the system (masses m_1 and m_2) remains unchanged:

$$\begin{aligned}
m_1\,\bar{v}_1 + m_2\,\bar{v}_2 &= \frac{1}{m_1 + m_2}\big[m_1^2\,v_1 + m_1\,m_2\,v_2 - e\,m_1\,m_2(v_1 - v_2) \\
&\quad + m_1\,m_2\,v_1 + m_2^2\,v_2 + e\,m_1\,m_2(v_1 - v_2)\big] \\
&= m_1\,v_1 + m_2\,v_2\,.
\end{aligned}$$

If we calculate the difference $\bar{v}_2 - \bar{v}_1$ of the velocities after the impact, we obtain

$$\bar{v}_2 - \bar{v}_1 = \frac{e\,(v_1 - v_2)(m_1 + m_2)}{m_1 + m_2} = e\,(v_1 - v_2)\,.$$

Here, $(v_1 - v_2)$ is the relative velocity of approach (just before the impact) of the masses, and $(\bar{v}_2 - \bar{v}_1)$ is the relative velocity of separation (immediately after impact). Thus, we have

$$e = -\frac{\bar{v}_1 - \bar{v}_2}{v_1 - v_2}\,. \tag{2.35}$$

Accordingly, the coefficient of restitution is equal to the ratio of the relative velocity of separation to the relative velocity of approach. In the following, we will usually apply (2.35) instead of (2.31).

The loss of mechanical energy (plastic deformation, generation of heat) during impact is given by the difference ΔT of the kinetic energies before and after impact. Applying (2.34) we obtain

$$\Delta T = \left(\frac{m_1 \, v_1^2}{2} + \frac{m_2 \, v_2^2}{2} \right) - \left(\frac{m_1 \, \bar{v}_1^2}{2} + \frac{m_2 \, \bar{v}_2^2}{2} \right)$$

$$= \frac{1 - e^2}{2} \, \frac{m_1 \, m_2}{m_1 + m_2} \, (v_1 - v_2)^2 \, . \tag{2.36}$$

There is no loss of energy in the case of an elastic impact ($e = 1$), whereas ΔT attains a maximum value for a plastic impact ($e = 0$).

In certain applications, for example, during forging or while driving a pile into the ground, the mass m_2 is at rest before the impact ($v_2 = 0$). We define the *blow efficiency* η as the ratio of the energy loss ΔT (= work done to cause the deformation) to the applied energy $T = \frac{1}{2} m_1 \, v_1^2$. Then we obtain with (2.36)

$$\eta = \frac{\Delta T}{T} = (1 - e^2) \frac{m_2}{m_1 + m_2} = (1 - e^2) \frac{1}{1 + \dfrac{m_1}{m_2}} \, . \tag{2.37}$$

It is the aim of forging to deform bodies plastically. Here, the blow efficiency η should be large. This can be achieved using a small ratio m_1/m_2 (large mass m_2 of the anvil including the work piece). On the other hand, a pile or a nail should not deform during driving (small η). Therefore, m_1/m_2 should be sufficiently large (large mass m_1 of the hammer).

We now extend our investigation to analyse oblique central impact. For simplicity we restrict ourselves to plane problems (Fig. 2.11a). We assume that the surfaces of the masses are smooth (frictionless); rough surfaces will be considered in Section 3.3.3. Then the contact force $F(t)$ and hence also the linear impulse $\hat{F}$ have the direction of the line of impact (Fig. 2.11b). Using the coordinate system shown in Fig. 2.11a, the Impulse Law in the y-direction yields

$$m_1 \, \bar{v}_{1y} - m_1 \, v_{1y} = 0 \quad \rightarrow \quad \bar{v}_{1y} = v_{1y},$$

$$m_2 \, \bar{v}_{2y} - m_2 \, v_{2y} = 0 \quad \rightarrow \quad \bar{v}_{2y} = v_{2y} \, . \tag{2.38}$$

Thus, the components of the velocities perpendicular to the line of impact remain unchanged in the case of smooth surfaces.

The equations in the direction of the line of impact (the x-axis)

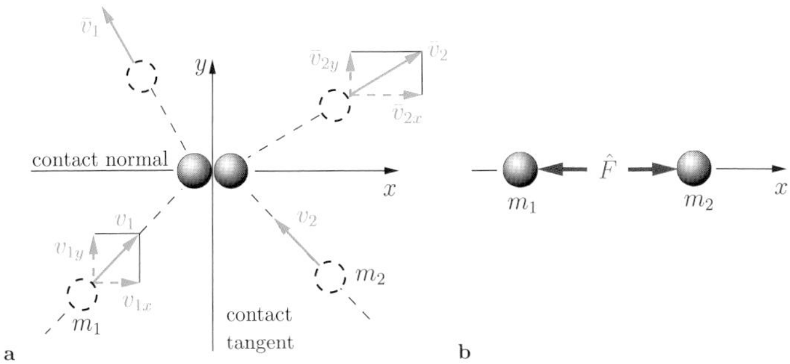

Fig. 2.11

and Equation (2.35) are the same as those of direct impact. Note, however, that the velocities of a direct collision have to be replaced now by the velocity components in the direction of the line of impact. In contrast to (2.32) and (2.33) we will write here the Impulse Laws for the total impact duration t_i:

$$
\begin{aligned}
m_1 \bar{v}_{1x} - m_1 v_{1x} &= -\hat{F}, \\
m_2 \bar{v}_{2x} - m_2 v_{2x} &= +\hat{F}.
\end{aligned}
\tag{2.39}
$$

Equation (2.35) now becomes

$$
e = -\frac{\bar{v}_{1x} - \bar{v}_{2x}}{v_{1x} - v_{2x}}.
\tag{2.40}
$$

These are three equations for the three unknowns $\bar{v}_{1x}$, $\bar{v}_{2x}$ and $\hat{F}$. Solving for $\bar{v}_{1x}$ and $\bar{v}_{2x}$ yields the results which are already known from (2.34).

Example 2.5 Two masses ($m_1 = m$, $m_2 = 2\,m$) collide in a straight path (Fig. 2.12). The velocity v_1 of m_1 is given.

Determine v_2 so that m_1 is at rest after the collision (coefficient of restitution e). Calculate the velocity of m_2 after the impact.

E2.5

Fig. 2.12

Solution The velocities after the collision in a direct central impact are given by (2.34). We assume that positive velocities are directed to the right. Then we obtain (note the direction of v_2)

$$\bar{v}_1 = \frac{m_1 v_1 - m_2 v_2 - e m_2 (v_1 + v_2)}{m_1 + m_2},$$

$$\bar{v}_2 = \frac{m_1 v_1 - m_2 v_2 + e m_1 (v_1 + v_2)}{m_1 + m_2}.$$

The condition $\bar{v}_1 = 0$ leads to

$$m_1 v_1 - m_2 v_2 - e m_2(v_1 + v_2) = 0$$

$$\rightarrow \quad \underline{\underline{v_2 = v_1 \frac{m_1 - e m_2}{m_2(1 + e)}}} = v_1 \frac{1 - 2 e}{2 (1 + e)}.$$

Inserting v_2 into $\bar{v}_2$ yields

$$\underline{\underline{\bar{v}_2}} = \frac{1}{3 m} \left[m v_1 - 2 m v_1 \frac{1 - 2 e}{2 (1 + e)} + e m \left(v_1 + v_1 \frac{1 - 2 e}{2 (1 + e)} \right) \right]$$

$$= \underline{\underline{v_1 \frac{3 e}{2 (1 + e)}}}.$$

Mass m_2 has to be at rest before the collision for $e = 1/2$. If $e > 1/2$, the direction of v_2 is reversed. In this case, mass m_2 has to move to the right before impact.

E2.6

Example 2.6 A mass m_1 slides down a smooth path from rest at point A. It collides with a mass $m_2 = 3 m_1$ which is at rest at point B (Fig. 2.13). The path is horizontal at B.

Determine the coefficients of restitution e which lead to a motion where mass m_1 moves uphill after the collision. Calculate the height h^* which is attained by m_1 for $e = 1/2$ and determine the travel distance w of m_2.

Solution The velocities of m_1 and m_2 immediately before impact are $v_1 = \sqrt{2 g h}$ (conservation of energy) and $v_2 = 0$. Equation

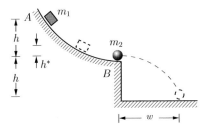

Fig. 2.13

(2.34) yields the velocities after impact:

$$\bar{v}_1 = \frac{m_1 - e\,m_2}{m_1 + m_2}\,v_1 = \frac{1 - 3\,e}{4}\,\sqrt{2\,gh},$$

$$\bar{v}_2 = \frac{m_1(1 + e)}{m_1 + m_2}\,v_1 = \frac{1 + e}{4}\,\sqrt{2\,gh}\,.$$

If m_1 is supposed to move uphill, the velocity $\bar{v}_1$ has to be negative. Therefore, the coefficient of restitution has to satisfy the condition

$$1 - 3\,e < 0 \quad \rightarrow \quad \underline{\underline{e > \frac{1}{3}}}\,.$$

In the special case $e = 1/2$ we obtain

$$\bar{v}_1 = -\frac{1}{8}\sqrt{2\,gh}, \quad \bar{v}_2 = \frac{3}{8}\sqrt{2\,gh}\,.$$

The height h^* is found from the conservation of energy:

$$\frac{1}{2}\,m_1\,\bar{v}_1^2 = m_1\,gh^* \quad \rightarrow \quad \underline{\underline{h^* = \frac{\bar{v}_1^2}{2\,g} = \frac{h}{64}}}\,.$$

The distance w follows from (1.41) with $\alpha = 0$ and $z(x = w) = -h$:

$$\underline{\underline{w}} = \bar{v}_2\sqrt{\frac{2\,h}{g}} = \underline{\underline{\frac{3}{4}\,h}}\,.$$

Example 2.7 A mass m_1 (velocity v_1) collides with a mass m_2 which is at rest. The plane of contact has the direction given by the angle $45°$ (Fig. 2.14a). The surfaces of the bodies are smooth.

Determine the velocities of the masses after the impact (coefficient of restitution e).

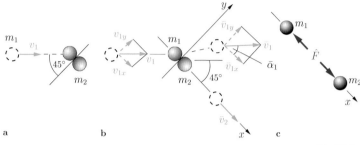

Fig. 2.14

Solution We choose the coordinate system shown in Fig. 2.14b; the x-axis is the line of impact. Since the surfaces are smooth, the linear impulse $\hat{F}$ acts in the direction of the line of impact (Fig. 2.14c). The Impulse Laws for both masses and the hypothesis (2.35) are given by

$$m_1(\bar{v}_{1x} - v_{1x}) = -\hat{F}, \quad m_1(\bar{v}_{1y} - v_{1y}) = 0,$$

$$m_2(\bar{v}_{2x} - v_{2x}) = +\hat{F}, \quad m_2(\bar{v}_{2y} - v_{2y}) = 0,$$

$$e = -\frac{\bar{v}_{1x} - \bar{v}_{2x}}{v_{1x} - v_{2x}}.$$

With

$$v_{1x} = v_{1y} = \frac{\sqrt{2}}{2} v_1, \quad v_{2x} = v_{2y} = 0$$

we obtain the velocities after the impact:

$$\underline{\bar{v}_{1x} = \frac{\sqrt{2}}{2} v_1 \frac{m_1 - e\, m_2}{m_1 + m_2}}, \quad \underline{\bar{v}_{1y} = \frac{\sqrt{2}}{2} v_1},$$

$$\underline{\bar{v}_{2x} = \frac{\sqrt{2}}{2} v_1 \frac{m_1(1 + e)}{m_1 + m_2}}, \quad \underline{\bar{v}_{2y} = 0}.$$

Thus, mass m_2 moves with the velocity $\bar{v}_2 = \bar{v}_{2x}$ in the direction of the line of impact (Fig. 2.14b). The velocity $\bar{v}_1$ and the angle $\bar{\alpha}_1$ are given by

$$\underline{\underline{\bar{v}_1 = \sqrt{\bar{v}_{1x}^2 + \bar{v}_{1y}^2}}} = \frac{v_1}{m_1 + m_2}\sqrt{m_1^2 + (1-e)m_1\,m_2 + \frac{1}{2}(1+e^2)\,m_2^2}\,,$$

$$\underline{\underline{\tan\bar{\alpha}_1 = \frac{\bar{v}_{1y}}{\bar{v}_{1x}}}} = \frac{m_1 + m_2}{m_1 - e\,m_2}\,.$$

2.6 Bodies with Variable Mass

Up to now we have always assumed that the mass of a system is constant. We will now extend the theory to systems which have a variable mass. An example for the motion of such a system is the flight of a rocket whose mass decreases with time.

Let us first consider a body B which initially has mass m_0 and velocity $\boldsymbol{v}_0$ (Fig. 2.15). At a certain time a mass Δm is ejected from B with the velocity $\boldsymbol{w}$. Then the mass of the body is reduced to $m_0 - \Delta m$ and the velocity is changed to $\boldsymbol{v}_1 = \boldsymbol{v}_0 + \Delta\boldsymbol{v}$. The *mass flow velocity* $\boldsymbol{w}$ is the velocity of Δm relative to the body *after* the ejection. The mass Δm therefore has the absolute velocity $\boldsymbol{v}_1 + \boldsymbol{w}$ (cf. Chapter 6). We assume that the system under consideration consists of both masses. Then, only *internal* forces act during the ejection.

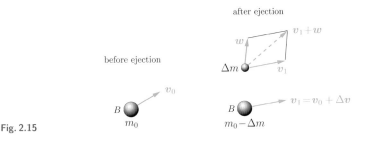

Fig. 2.15

The momentum of the system before the ejection

$$\boldsymbol{p}_0 = m_0\,\boldsymbol{v}_0$$

and the momentum afterwards

$$\boldsymbol{p}_1 = (m_0 - \Delta m)\boldsymbol{v}_1 + \Delta m(\boldsymbol{v}_1 + \boldsymbol{w})$$

have to be equal according to (2.14): $\boldsymbol{p}_0 = \boldsymbol{p}_1$. This yields the change in the velocity of the body B due to the ejected mass:

$$\Delta\boldsymbol{v} = \boldsymbol{v}_1 - \boldsymbol{v}_0 = -\frac{\Delta m}{m_0}\,\boldsymbol{w}\,. \tag{2.41}$$

This change increases with an increasing mass Δm and an increasing velocity $\boldsymbol{w}$. The negative sign in (2.41) indicates that $\Delta\boldsymbol{v}$ and $\boldsymbol{w}$ are oppositely directed. If a mass Δm *hits* the body B with a relative velocity $\boldsymbol{w}$ and is *absorbed* by B (the body gains mass instead of losing mass), then the algebraic sign in (2.41) is reversed.

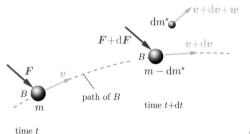

Fig. 2.16

Consider now a body B which ejects mass continuously and which is subjected to an external force $\boldsymbol{F}$ (Fig. 2.16). The body has mass m and velocity $\boldsymbol{v}$ at time t. During a time interval dt it ejects mass dm^* with a mass flow velocity $\boldsymbol{w}$. At time $t + dt$ its mass is therefore $m - dm^*$; its velocity has been changed by $d\boldsymbol{v}$. The momentum of the system at time t is given by

$$\boldsymbol{p}(t) = m\,\boldsymbol{v}$$

and at time $t + \mathrm{d}t$ it is

$$\boldsymbol{p}(t + \mathrm{d}t) = (m - \mathrm{d}m^*)(\boldsymbol{v} + \mathrm{d}\boldsymbol{v}) + \mathrm{d}m^*(\boldsymbol{v} + \mathrm{d}\boldsymbol{v} + \boldsymbol{w})$$
$$= m\,\boldsymbol{v} + m\,\mathrm{d}\boldsymbol{v} + \mathrm{d}m^*\,\boldsymbol{w} = \boldsymbol{p}(t) + \mathrm{d}\boldsymbol{p}\,.$$

Thus, (2.12) yields

$$\frac{\mathrm{d}\boldsymbol{p}}{\mathrm{d}t} = m\,\frac{\mathrm{d}\boldsymbol{v}}{\mathrm{d}t} + \frac{\mathrm{d}m^*}{\mathrm{d}t}\,\boldsymbol{w} = \boldsymbol{F}, \tag{2.42}$$

where $\mathrm{d}m^*/\mathrm{d}t = \mu$ is the mass which is *ejected* per unit time (the rate at which mass is being ejected). The rate of *change of the mass* $\mathrm{d}m/\mathrm{d}t$ of the body is given by $-\mathrm{d}m^*/\mathrm{d}t$ (rate of mass loss):

$$\frac{\mathrm{d}m}{\mathrm{d}t} = -\,\frac{\mathrm{d}m^*}{\mathrm{d}t} = -\,\mu\,. \tag{2.43}$$

Now we introduce the *thrust* $\boldsymbol{T}$:

$$\boldsymbol{T} = -\,\mu\boldsymbol{w}\,. \tag{2.44}$$

Then (2.42) can be written as

$$m\,\frac{\mathrm{d}\boldsymbol{v}}{\mathrm{d}t} = \boldsymbol{F} + \boldsymbol{T}\,. \tag{2.45}$$

This equation has the same form as Newton's law of motion. Note, however, that the mass of the body now depends on time: $m = m(t)$. Also, it contains the thrust $\boldsymbol{T}$ in addition to the external force $\boldsymbol{F}$. The thrust describes the action of the ejected mass on the body. It is proportional to the ejected mass μ and to the mass flow velocity $\boldsymbol{w}$; it acts on the body in the direction opposite to $\boldsymbol{w}$. For example, if a rocket expels mass backwards, the thrust acts on the rocket forwards. The thrust increases with increasing mass flow velocity $\boldsymbol{w}$.

As an illustrative example let us consider a rocket which has the initial mass m_I (including the fuel); see Fig. 2.17a. It takes off vertically from the surface of the earth with a constant thrust and a constant rate of expelled mass. The forces acting on the rocket are the thrust T (directed opposite to the velocity w) and the time-dependent weight $m(t)\,g$ (Fig. 2.17b). We neglect the

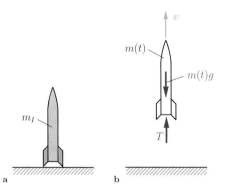

Fig. 2.17

a   b

aerodynamic drag and assume g to be constant. Then the motion of the rocket is described according to (2.45) by

$$m(t)\,\dot{v} = -m(t)\,g + T$$

where

$$T = \mu w = -\dot{m}w\,.$$

The condition for lift off is $\dot{v}(0) > 0$. Therefore, the thrust has to satisfy $T > m_I\,g$. Inserting yields

$$\frac{dv}{dt} = -g - \frac{1}{m}\frac{dm}{dt}\,w\,.$$

Since T and μ are assumed constant, we also have $w = \text{const}$. Thus, integration and application of the initial condition $v(0) = 0$ yields the velocity

$$v(t) = -gt - w \int\limits_{m_I}^{m(t)} \frac{d\bar{m}}{\bar{m}} = -gt - w \ln\frac{m(t)}{m_I} = w \ln\frac{m_I}{m(t)} - gt\,.$$

From $\dot{m} = -\mu$ we obtain $m(t) = m_I - \mu t$ which leads to

$$v(t) = w \ln\frac{m_I}{m_I - \mu t} - gt\,.$$

The maximum velocity is reached at the time $t = t_F$ when the fuel has run out. With the final mass $m(t_F) = m_F$ it follows as

$$v_{\max} = w \ln \frac{m_I}{m_F} - g t_F .$$

It increases with increasing w and increasing ratio m_I/m_F.

Example 2.8 A boat (total mass m_0) at rest can move without resistance in the water. Two masses m_1 and m_2 are thrown off the boat in the same direction with a mass flow velocity w.

E2.8

   Determine the resulting velocities of the boat if
a) the two masses are thrown at the same time and
b) if mass m_1 is thrown first and subsequently mass m_2.

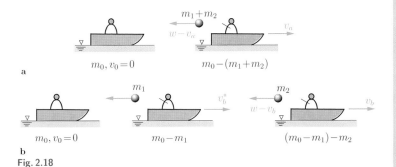

Fig. 2.18

Solution Fig. 2.18a shows the situations before and after the ejection of the masses in case a). The ejected mass $m_1 + m_2$ moves relative to the boat with the velocity w to the left. If the boat moves with the velocity v_a to the right, then the masses $m_1 + m_2$ have the absolute velocity $w - v_a$ to the left. Since the initial momentum is zero ($v_0 = 0$), the principle of conservation of linear momentum for the total system is given by

$$(m_0 - m_1 - m_2)\, v_a - (m_1 + m_2)(w - v_a) = 0 .$$

Solving for the velocity v_a of the boat after the ejection yields

$$v_a = \frac{m_1 + m_2}{m_0}\, w .$$

In case b) the velocity v_b^* of the boat after the ejection of the first mass m_1 (Fig. 2.18b) is obtained as

$$(m_0 - m_1)\, v_b^* - m_1(w - v_b^*) = 0 \quad \rightarrow \quad v_b^* = \frac{m_1}{m_0}\, w\,.$$

Application of the principle of conservation of linear momentum to the system $(m_0 - m_1)$ before and after the ejection of the second mass m_2 leads to

$$(m_0 - m_1)\, v_b^* = (m_0 - m_1 - m_2)\, v_b - m_2(w - v_b)$$

$$\rightarrow \quad v_b = \left(\frac{m_1}{m_0} + \frac{m_2}{m_0 - m_1} \right) w\,.$$

After simple algebraic manipulation the velocity v_b can be written as

$$v_b = \left(\frac{m_1 + m_2}{m_0} + \frac{m_1\, m_2}{m_0(m_0 - m_1)} \right) w = v_a + \frac{m_1\, m_2}{m_0(m_0 - m_1)}\, w\,.$$

Since $m_0 > m_1$, the velocity of the boat in case b) is larger than the velocity in case a).

The results for v_a and v_b can also be obtained by a repeated application of (2.41).

E2.9

Example 2.9   The end of a chain with mass m_0 and length l is pulled upwards with a constant acceleration a_0 (Fig. 2.19a). Determine the necessary force H.

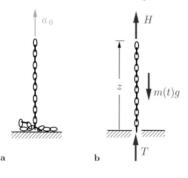

Fig. 2.19

Solution   We consider the part of the chain which is already suspended to be a body whose mass is continuously increasing. This body is subjected to the force H, the time-dependent weight $m(t)\,g$ and the "thrust" T. We assume the force T to be acting upwards (Fig. 2.19b). With the z-coordinate of the point of application of H as shown, Equation (2.45) yields

$$m(t)\ddot{z} = H - m(t)\,g + T\,,\tag{a}$$

where the thrust is given according to (2.43) and (2.44):

$$T = \dot{m}\,w\,.\tag{b}$$

The part of the chain which is still at rest has a "velocity" in the negative z-direction relative to the part moving with the velocity $\dot{z}$. Thus,

$$w = -\dot{z}\,.\tag{c}$$

Integration of the given acceleration a_0 and application of the initial conditions $\dot{z}(0) = 0$ and $z(0) = 0$ leads to the velocity and the position ($=$ length of the moving part):

$$\ddot{z} = a_0, \quad \dot{z} = a_0\,t, \quad z = \frac{1}{2}\,a_0\,t^2\,.\tag{d}$$

This yields the mass of the body and its change of mass:

$$m = m_0\frac{z}{l} = \frac{m_0\,a_0}{2\,l}\,t^2,\tag{e}$$

$$\dot{m} = \frac{m_0\,a_0}{l}\,t\,.\tag{f}$$

If we introduce (b)-(f) into (a) and solve for H we obtain

$$\underline{\underline{H = \frac{m_0\,a_0(3\,a_0 + g)}{2\,l}\,t^2 = m_0(3\,a_0 + g)\frac{z}{l}\,.}}$$

This result is valid only as long as the mass of the body changes ($z < l$).

2.7 Supplementary Examples

Detailed solutions to the following examples are given in (**A**) D. Gross et al. *Formeln und Aufgaben zur Technischen Mechanik 3*, Springer, Berlin 2010, or (**B**) W. Hauger et al. *Aufgaben zur Technischen Mechanik 1-3*, Springer, Berlin 2008.

E2.10

Example 2.10 Two vehicles (masses m_1 and m_2, velocities v_1 and v_2) crash head-on, see Fig. 2.20. After a plastic impact the vehicles are entangled and slide with locked wheels a distance s to the right. The coefficient of kinetic friction between the wheels and the road is μ.

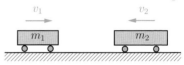

Calculate v_1 if v_2 and s are known.

Fig. 2.20

Result: see (**B**) $v_1 = \dfrac{m_2}{m_1} v_2 + \left(1 + \dfrac{m_2}{m_1}\right) \sqrt{2\,\mu\,g\,s}$.

E2.11

Example 2.11 A block (mass m_2) rests on a horizontal platform (mass m_1) which is also initially at rest (Fig. 2.21). A constant force F accelerates the platform (wheels rolling without friction) which causes the block to slide on the rough surface of the platform (coefficient of kinetic friction μ).

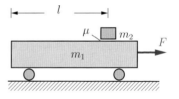

Fig. 2.21

Determine the time t^* that it takes the block to fall off the platform.

Result: see (**A**) $t^* = \sqrt{\dfrac{2lm_1}{F - \mu g\,(m_1 + m_2)}}$.

E2.12

Example 2.12 A railroad wagon (mass m_1) has a velocity v_1 (Fig. 2.22). It collides with a wagon (mass m_2) which is initially at rest. Both wagons roll without friction after the collision. The second wagon is connected via a spring (spring constant k) with a block (mass m_3) that lies on a rough surface (coefficient of static friction μ_0).

Assume the impact to be plastic and determine the maximum value of v_1 so that the block stays at rest.

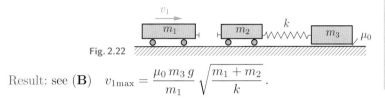

Fig. 2.22

Result: see (**B**) $v_{1\text{max}} = \dfrac{\mu_0 \, m_3 \, g}{m_1} \sqrt{\dfrac{m_1 + m_2}{k}}$.

Example 2.13   A point mass m_1 strikes a point mass m_2 which is suspended from a string (length l, negligible mass) as shown in Fig. 2.23. The maximum force S^* that the string can sustain is given.

   Assume an elastic impact and determine the velocity v_0 that causes the string to break.

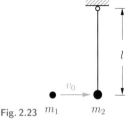

Fig. 2.23 m_1 m_2

Result: see (**B**) $v_0 > \dfrac{m_1 + m_2}{2 \, m_1} \sqrt{l \left(S^*/m_2 - g \right)}$.

Example 2.14   A ball ① (mass m_1) hits a second ball ② (mass m_2, velocity $v_2 = 0$) with a velocity v_1 as shown in Fig. 2.24. Assume that the impact is partially elastic (coefficient of restitution e) and all surfaces are smooth. Given: $r_2 = 3 \, r_1$, $m_2 = 4 \, m_1$.

   Determine the velocities of the balls after the collision.

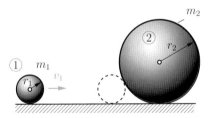

Fig. 2.24

Results: see (**A**) $\bar{v}_1 = \dfrac{1 - 3e}{4} \, v_1$, $\bar{v}_2 = \dfrac{\sqrt{3}}{8}(1 + e) \, v_1$.

Example 2.15   A hunter (mass m_1) sits in a boat (mass $m_2 = 2 \, m_1$) which can move in the water without resistance. The boat is initially at rest.

a) Determine the velocity v_{B_1} of the boat after the hunter fires a bullet (mass $m_3 = m_1/1000$) with a velocity $v_0 = 500$ m/s.

b) Find the direction of the velocity of the boat after a second
shot is fired at an angle of $45°$ with respect to the first one.

Results: see (**A**)   a) $v_{B_1} = 0.167\,\text{m/s}$,   b) $\alpha = 22.5°$.

E2.16   **Example 2.16**   A car ② goes into a skid on a wet road and comes to a stop sideways across the road as shown in Fig. 2.25. In
spite of having fully applied the brakes a distance s_1 from car ②
a second car ① (sliding with the coeffi-
cient of kinetic friction μ) collides with
car ②. This causes car ② to slide a distance s_2. Assume a partially elastic central
impact. Given: $m_1 = 2\,m_2$, $\mu = 1/3$, $e =$
0.2, $s_1 = 50\,\text{m}$, $s_2 = 10\,\text{m}$.

Determine the velocity v_0 of car ①
before the breaks were applied.

Result: see (**A**) $v_0 = 74.6\,\text{km/h}$.

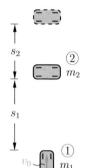

Fig. 2.25

E2.17   **Example 2.17**   Two cars (point masses m_1 and m_2) collide at an
intersection with the velocities v_1 and v_2 at an angle α (Fig. 2.26).
Assume a perfectly plastic
collision.

Determine the magnitu-
de and the direction of the
velocity immediately after
the impact. Calculate the
loss of energy during the collision.
Results: see (**A**)

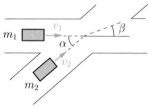

Fig. 2.26

$$\bar{v} = \frac{1}{m_1 + m_2}\sqrt{(m_1 v_1)^2 + 2m_1 m_2 v_1 v_2 \cos\alpha + (m_2 v_2)^2}\,,$$

$$\tan\beta = \frac{m_2 v_2 \sin\alpha}{m_1 v_1 + m_2 v_2 \cos\alpha}\,,$$

$$\Delta T = \frac{m_1 m_2}{2(m_1 + m_2)}(v_1^2 + v_2^2 - 2v_1 v_2 \cos\alpha)\,.$$

Example 2.18 A bullet (mass m) has a velocity v_0 (Fig. 2.27). An explosion causes the bullet to break into two parts (point masses m_1 and m_2). The directions α_1 and α_2 of the two parts and the velocity v_1 immediately after the explosion are given.

E2.18

Calculate m_1 and v_2. Determine the trajectory of the center of mass of the two parts.

Results: see (**A**)

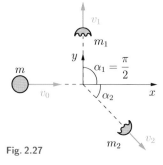

Fig. 2.27

$$m_1 = m \frac{v_0}{v_1} \tan \alpha_2 ,$$

$$v_2 = \frac{v_0}{\cos \alpha_2 - \frac{v_0}{v_1} \sin \alpha_2} , \quad y_c = 0 .$$

Example 2.19 A ball (point mass m_1) is attached to a cable. It is released from rest at the height h_1 (Fig. 2.28). After falling to the vertical position at A it collides with a second ball (point mass $m_2 = 2\,m_1$) which is also initially at rest. The coefficient of restitution is $e = 0.8$.

E2.19

Determine the height h_2 which the first ball can reach after the collision and the velocity of the second ball immediately after impact.

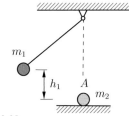

Fig. 2.28

Results: see (**A**) $h_2 = 0.04\, h_1 ,$ $\bar{v}_2 = 0.6\,\sqrt{2 g h_1} .$

Example 2.20 The rigid rod (negligible mass) in Fig. 2.29 carries two point masses. It is struck by an impulsive force $\hat{F}$ at a distance a from the support A.

E2.20

Determine the angular velocity of the rod immediately after the impact and the impulsive reaction at A. Calculate a so that the reaction force at A is zero.

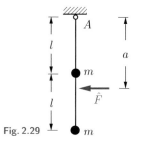

Fig. 2.29

Results: see (**B**) $\bar{\omega} = \dfrac{\hat{F}\,a}{5\,l^2\,m} ,$ $\hat{A}_x = \left(1 - \dfrac{3a}{5\,l}\right)\hat{F} ,$ $a = 5\,l/3 .$

E2.21

Example 2.21 A wagon (weight $W_1 = m_1 g$) hits a buffer (spring constant k) with the velocity v. The collision causes a block (weight $W_2 = m_2 g$) to slide on the rough surface (coefficient of static friction μ_0) of the wagon (Fig. 2.30).

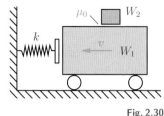

What is the least speed the wagon was doing before the collision?

Result: see (**A**) $v = \mu_0 g \sqrt{\dfrac{m_1 + m_2}{k}}$.

Fig. 2.30

E2.22

Example 2.22 The system shown in Fig. 2.31 consists of two blocks (masses m_1 and m_2), a spring (spring constant k), a massless rope and two massless pulleys. Block 1 lies on a rough surface (coefficient of kinetic friction μ).

At the beginning of the motion ($t = 0$), the spring is unstretched and the position of block 1 is given by $x_1 = 0$.

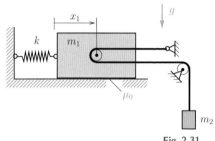

Determine the velocity $\dot{x}_1$ as a function of the position x_1.

Fig. 2.31

Result: $\dot{x}_1 = \sqrt{\dfrac{2}{5}(2 - \mu)g x_1 - \dfrac{1}{5}\dfrac{k}{m}x_1^2}$.

2.8 Summary

- Determination of the motion of the individual point masses:
  - ◇ Derivation of the equations of motion.
  - ◇ Formulation of the kinematic equations.

- Law of motion for the center of mass: $m\,\boldsymbol{a}_c = \boldsymbol{F}$,

  $\boldsymbol{a}_c$ acceleration of the center of mass of the system,

  $\boldsymbol{F}$ resultant of the external forces.

- Conservation of linear momentum:

  $\boldsymbol{p} = m\,\boldsymbol{v}_c = \sum_i m_i \boldsymbol{v}_i = \text{const}$,

  $\boldsymbol{v}_c$ velocity of the center of mass of the system.

  Note: no external forces are acting.

- Angular momentum theorem: $\dot{\boldsymbol{L}}^{(0)} = \boldsymbol{M}^{(0)}$,

  $\boldsymbol{L}^{(0)} = \sum_i (\boldsymbol{r}_i \times m_i\,\boldsymbol{v}_i)$ moment of momentum.

- Work-energy theorem: $T_1 - T_0 = U^{(e)} + U^{(i)}$,

  in the case of rigid constraints $U^{(i)} = 0$.

- Conservation of energy: $T + V^{(e)} + V^{(i)} = \text{const}$,

  in the case of rigid constraints $T + V^{(e)} = \text{const}$.

- Impact problems:
  - ◇ Choice of a coordinate system: line of impact (x), tangent (y).
  - ◇ Application of the Impulse Law for each point mass.
  - ◇ Application of the hypothesis $e = -\dfrac{\bar{v}_{1x} - \bar{v}_{2x}}{v_{1x} - v_{2x}}$.

- Systems with variable mass: $m\,\boldsymbol{a} = \boldsymbol{F} + \boldsymbol{T}$,

  $\boldsymbol{T} = -\mu\,\boldsymbol{w} = \dot{m}\,\boldsymbol{w}$ thrust,

  $\boldsymbol{w}$ mass flow velocity.

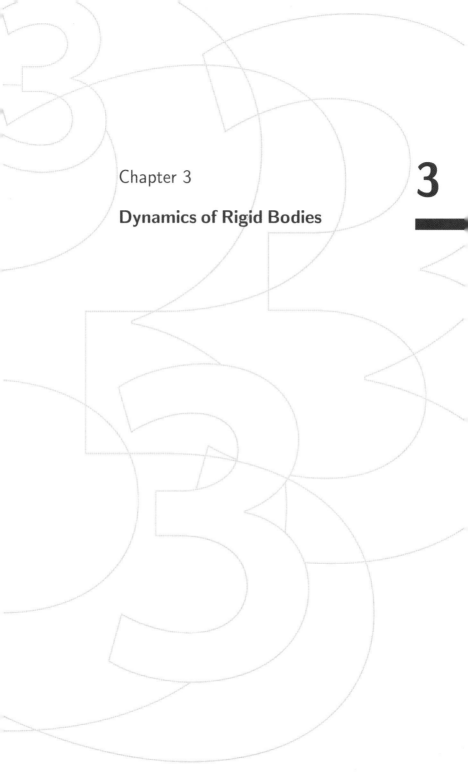

Chapter 3

Dynamics of Rigid Bodies

3

3 Dynamics of Rigid Bodies

———— Objectives: A rigid body may be considered to be a system of an infinite number of particles whose relative distances remain unchanged when the body is loaded. As was explained in Section 2.1, it has six degrees of freedom in space: three translations (in the x-, y-, and z-directions) and three rotations (about the x-, y-, and z-axes). In the following chapter we will derive the equations which describe the motion of rigid bodies and we will explain how these equations are applied to specific problems. Of particular interest will be plane motion and the rotation about a fixed axis.

3.1 Kinematics

In this section we will study the kinematics of a rigid body, i.e., the geometry of motion without reference to its cause.

A *rigid body* may be considered to be a system of an infinite number of particles whose relative distances remain unchanged when the body is loaded. It has six degrees of freedom in space (see Section 2.1). Three translations (in the x-, y-, and z-directions) and three rotations (about the x-, y-, and z-axes) correspond to the six degrees of freedom. In the following we will show how the general motion of a rigid body may be understood as the composition of a translation and a rotation.

3.1.1 Translation

A motion that leaves the direction of the straight line between any two arbitrary points A and P of a rigid body unchanged is called a *translation* (Fig. 3.1). In this case, every particle of the body undergoes the same displacement $\mathrm{d}\boldsymbol{r}$ during a time interval $\mathrm{d}t$. Therefore, all the particles have the same velocity and the same acceleration:

$$\boldsymbol{v} = \frac{\mathrm{d}\boldsymbol{r}}{\mathrm{d}t}, \quad \boldsymbol{a} = \frac{\mathrm{d}\boldsymbol{v}}{\mathrm{d}t} = \frac{\mathrm{d}^2\boldsymbol{r}}{\mathrm{d}t^2}. \tag{3.1}$$

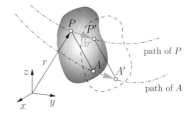

Fig. 3.1

The paths of different points all have the same shape. Thus, the motion of *a single point* of the body arbitrarily chosen represents the motion of the complete body.

3.1.2 Rotation

During a *rotation* all the particles of a rigid body move about a common axis. In the special case that this axis is fixed in space, the motion is called a *rotation about a fixed axis*. If on the other hand the axis only passes through a fixed point without keeping its direction, then the motion is referred to as a *rotation about a fixed point* or a *gyroscopic motion*.

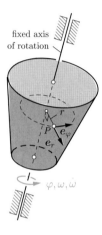

fixed axis
of rotation

Fig. 3.2

Let us first consider the motion of a rigid body about a fixed axis (Fig. 3.2). In this case each point of the body moves in a circle whose plane is perpendicular to the axis. The radius vectors from the axis to the individual points of the body sweep out the same angle $d\varphi$ during the same time interval dt. Thus, the angular velocity $\omega = \dot{\varphi}$ and the angular acceleration $\dot{\omega} = \ddot{\varphi}$, respectively, are the same for every point. The velocity and the acceleration of an arbitrary point P at a distance r from the axis are therefore the same as for a particle in a circular motion (see (1.25) - (1.28)):

$$\boldsymbol{v}_P = v_\varphi\,\boldsymbol{e}_\varphi, \quad \boldsymbol{a}_P = a_r\,\boldsymbol{e}_r + a_\varphi\,\boldsymbol{e}_\varphi \tag{3.2a}$$

where

$$v_\varphi = r\omega, \quad a_r = -r\omega^2, \quad a_\varphi = r\dot{\omega}. \tag{3.2b}$$

We now consider the rotation about a fixed point A (Fig. 3.3). Let the instantaneous direction of the axis of rotation be given by the unit vector $\boldsymbol{e}_\omega$. Assume that the body undergoes a rotation about this axis with the angle $\mathrm{d}\varphi$ during the time interval $\mathrm{d}t$. Then all the particles of the body *instantaneously* move in circles.

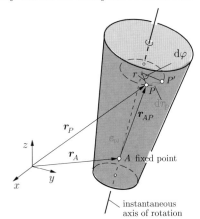

Fig. 3.3

The displacement $\mathrm{d}\boldsymbol{r}_P$ of an arbitrary point P is given by (see Fig. 3.3)

$$\mathrm{d}\boldsymbol{r}_P = (\boldsymbol{e}_\omega \times \boldsymbol{r}_{AP})\,\mathrm{d}\varphi . \tag{3.3}$$

Here, the vector $\boldsymbol{e}_\omega \times \boldsymbol{r}_{AP}$ is perpendicular to $\boldsymbol{e}_\omega$ and $\boldsymbol{r}_{AP}$; its magnitude is equal to the orthogonal distance r of point P from the instantaneous axis of rotation. We now introduce the *infinitesimal vector of rotation* $\mathrm{d}\boldsymbol{\varphi}$ and the *angular velocity vector* $\boldsymbol{\omega}$:

$$\mathrm{d}\boldsymbol{\varphi} = \mathrm{d}\varphi\,\boldsymbol{e}_\omega \quad \text{and} \quad \boldsymbol{\omega} = \frac{\mathrm{d}\boldsymbol{\varphi}}{\mathrm{d}t} = \dot\varphi\,\boldsymbol{e}_\omega = \omega\,\boldsymbol{e}_\omega . \tag{3.4}$$

Then the velocity $\boldsymbol{v}_P = \mathrm{d}\boldsymbol{r}_P/\mathrm{d}t$ of point P follows from (3.3):

$$\boldsymbol{v}_P = \boldsymbol{\omega} \times \boldsymbol{r}_{AP} . \tag{3.5}$$

It should be noted that the infinitesimal rotation $\mathrm{d}\boldsymbol{\varphi}$ and the angular velocity $\boldsymbol{\omega} = \mathrm{d}\boldsymbol{\varphi}/\mathrm{d}t$ are vectors, however, a finite rotation *cannot* be represented by a vector. In order to show this we subject a body to different finite rotations from an initial position to a final

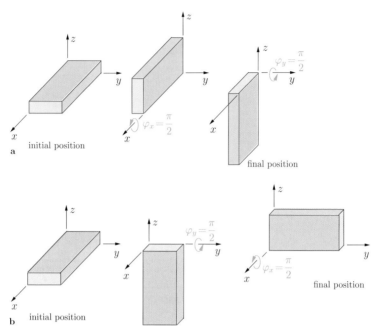

Fig. 3.4

position. For example, if the block in Fig. 3.4 is *first* rotated about the x-axis (angle of rotation $\varphi_x = \pi/2$) and *subsequently* rotated about the y-axis (angle of rotation $\varphi_y = \pi/2$), then the final position shown in Fig. 3.4a is attained. On the other hand, if we *first* rotate the body about the y-axis and *then* about the x-axis, we obtain a different final position, see Fig. 3.4b. According to the commutative law of vector addition, the result of an addition of vectors has to be independent of the sequence of the addition. Since finite angles of rotation do not obey this law, they cannot be classified as vectors.

The acceleration of P is obtained through differentiation of (3.5):

$$\boldsymbol{a}_P = \frac{\mathrm{d}\boldsymbol{v}_P}{\mathrm{d}t} = \dot{\boldsymbol{\omega}} \times \boldsymbol{r}_{AP} + \boldsymbol{\omega} \times \dot{\boldsymbol{r}}_{AP}.$$

Since point A is fixed in space ($\dot{\boldsymbol{r}}_A \equiv \boldsymbol{0}$), we have $\dot{\boldsymbol{r}}_{AP} = \dot{\boldsymbol{r}}_P = $

$= \boldsymbol{v}_P = \boldsymbol{\omega} \times \boldsymbol{r}_{AP}$. Thus,

$$\boldsymbol{a}_P = \dot{\boldsymbol{\omega}} \times \boldsymbol{r}_{AP} + \boldsymbol{\omega} \times (\boldsymbol{\omega} \times \boldsymbol{r}_{AP}). \tag{3.6}$$

Equations (3.5) and (3.6) reduce to (3.2a,b) in the special case of a rotation about a fixed axis.

3.1.3 General Motion

The general motion of a rigid body can be understood as a composition of a rotation and a translation. To show this, we first consider the case of *plane motion*, where all the particles move in the x, y-plane or in a plane parallel to it (Fig. 3.5a). Position vectors to arbitrary points P and A which are fixed in the body are connected by $\boldsymbol{r}_P = \boldsymbol{r}_A + \boldsymbol{r}_{AP}$. Let us introduce the unit vectors $\boldsymbol{e}_r$ (in the direction from A to P) and $\boldsymbol{e}_\varphi$ (perpendicular to $\boldsymbol{r}_{AP}$).

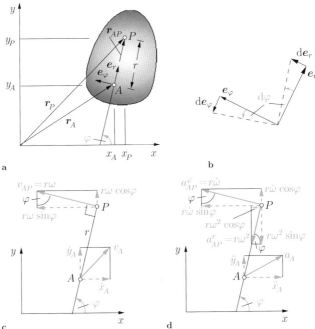

Fig. 3.5

They are also fixed in the body and therefore move with the body. Since $r_{AP} = r\, e_r$, we can write

$$r_P = r_A + r\, e_r.$$

Note that $r = \text{const}$. Therefore, differentiation yields $\dot{r}_P = \dot{r}_A + r\dot{e}_r$. The vector $\dot{e}_r$ is found through the following considerations. If the vector r_{AP} rotates through the angle $d\varphi$ during the infinitesimal time interval dt, then the vectors e_r and e_φ are also rotated by $d\varphi$. According to Fig. 3.5b we have $de_r = d\varphi\, e_\varphi$ and therefore $\dot{e}_r = de_r/dt = \dot{\varphi}\, e_\varphi$. Similarly, we obtain $\dot{e}_\varphi = -\dot{\varphi}\, e_r$ (cf. Section 1.1.4). Thus, the velocity of P is given by

$$\dot{r}_P = \dot{r}_A + r\omega\, e_\varphi$$

where $\omega = \dot{\varphi}$, and the acceleration is found to be

$$\ddot{r}_P = \ddot{r}_A + r\dot{\omega}\, e_\varphi + r\omega\, \dot{e}_\varphi = \ddot{r}_A + r\dot{\omega}\, e_\varphi - r\omega^2 e_r \,.$$

In summary we have

$$r_P = r_A + r_{AP},$$

$$v_P = v_A + v_{AP}, \qquad (3.7a)$$

$$a_P = a_A + a_{AP}^r + a_{AP}^\varphi$$

with

$$r_{AP} = r\, e_r, \quad v_{AP} = r\omega\, e_\varphi,$$

$$a_{AP}^r = -r\omega^2 e_r, \quad a_{AP}^\varphi = r\dot{\omega}\, e_\varphi \,. \qquad (3.7b)$$

Each of the relations (3.7a) consists of two parts. The quantities r_A, v_A and a_A represent the translation of the body, whereas the other terms (see (3.7b)) represent a rotation (circular motion of P about A, cf. (3.2a,b)). The vectors v_{AP} and a_{AP}^φ are perpendicular to r_{AP}; the vector a_{AP}^r points in the direction from P to A (centripetal acceleration). Thus, the velocity (acceleration) of an arbitrary point P is equal to the sum of the velocity (acceleration) of an arbitrary point A and the velocity (acceleration) of point P due to the rotation about A.

Frequently we have to express the velocity and acceleration of P in a Cartesian coordinate system. To do so, we first write down the coordinates of P (see Fig. 3.5a):

$$x_P = x_A + r\cos\varphi, \quad y_P = y_A + r\sin\varphi.$$

If we differentiate once ($\varphi = \varphi(t)$; chain rule!) we obtain the components of the velocity vector and a second time those of the acceleration vector ($\dot{\varphi} = \omega$):

$$v_{Px} = \dot{x}_P = \dot{x}_A - r\omega\sin\varphi,$$
$$v_{Py} = \dot{y}_P = \dot{y}_A + r\omega\cos\varphi,$$
$$a_{Px} = \ddot{x}_P = \ddot{x}_A - r\dot{\omega}\sin\varphi - r\omega^2\cos\varphi,$$
$$a_{Py} = \ddot{y}_P = \ddot{y}_A + r\dot{\omega}\cos\varphi - r\omega^2\sin\varphi.$$

The meaning of the individual terms can be seen in Figs. 3.5c,d.

Equations (3.7a,b) can be used to determine the velocity (acceleration) of point P graphically with the aid of a *velocity diagram* *(acceleration diagram)*. The directions of the individual velocity vectors (acceleration vectors) have to be taken from a *layout diagram* which represents the geometrical properties of the problem. Consider, for example, the body in Fig. 3.6a. We assume that v_A, a_A, ω and $\dot{\omega}$ are known in the position shown, i.e., at a given instant. According to (3.7a) the velocity v_P is the sum of the vectors v_A and v_{AP} (Fig. 3.6b). The vector v_{AP} has the magnitude $v_{AP} = r\omega$ (see (3.7b)) and it is orthogonal to $\overline{AP}$.

The acceleration a_P (Fig. 3.6c) is obtained as the sum of a_A, a_{AP}^r (in the direction from P to A, $a_{AP}^r = r\omega^2$) and a_{AP}^φ (orthogonal to $\overline{AP}$, $a_{AP}^\varphi = r\dot{\omega}$). In the case of a purely graphical solution, the velocity diagram and the acceleration diagram have to be drawn to scale. If we want to solve the problem with a mixed graphical/analytical method (see Volume 1, Section 2.4) it suffices to sketch the diagrams but not necessarily to scale and apply trigonometry.

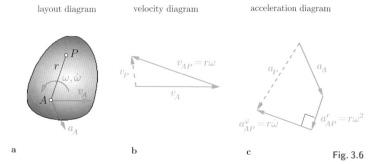

layout diagram   velocity diagram   acceleration diagram

a   b   c   **Fig. 3.6**

As an example, consider the rod in Fig. 3.7a. The velocity v_A and acceleration a_A of point A in the position shown are given; let us find the velocity v_B and acceleration a_B of point B at this instant. We first solve the problem analytically by introducing an x, y-coordinate system and the angle φ as shown in Fig. 3.7b. Note that the horizontal displacement of B is zero. The coordinates of B can be expressed as

$$x_B = x_A - l \sin \varphi = 0, \quad y_B = l \cos \varphi.$$

Differentiating yields $(\dot{x}_A = v_A)$

$$\dot{x}_B = v_A - l \dot{\varphi} \cos \varphi = 0 \quad \rightarrow \quad \dot{\varphi} = \omega = \frac{v_A}{l \cos \varphi},$$

$$v_B = \dot{y}_B = -l \omega \sin \varphi = -v_A \tan \varphi.$$

We differentiate again and use $\dot{v}_A = a_A$ to obtain

$$\ddot{x}_B = a_A - l \dot{\omega} \cos \varphi + l \omega^2 \sin \varphi = 0$$

$$\rightarrow \quad \dot{\omega} = \frac{a_A}{l \cos \varphi} + \frac{v_A^2 \sin \varphi}{l^2 \cos^3 \varphi},$$

$$a_B = \ddot{y}_B = -l \dot{\omega} \sin \varphi - l \omega^2 \cos \varphi = -a_A \tan \varphi - \frac{v_A^2}{l \cos^3 \varphi}.$$

The problem can also be solved with a mixed graphical/analytical method. The magnitude and the direction (horizontal) of the velocity v_A are known. We also know the direction of v_{AB} (orthogonal to $\overline{AB}$, $v_{AB} = l \omega$) and the direction of v_B (vertical). This enables us to sketch the velocity diagram (Fig. 3.7c) from which we obtain

$$l\omega = \frac{v_A}{\cos\varphi}, \quad v_B = v_A \tan\varphi.$$

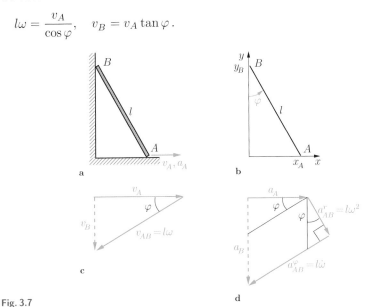

Fig. 3.7

To construct the acceleration diagram (Fig. 3.7d) we start with the given acceleration a_A (horizontal). We also know a_{AB}^r (in the direction from B to A, $a_{AB}^r = l\omega^2$) and the directions of a_{AB}^φ (orthogonal to $\overline{AB}$) and a_B (vertical). Now the acceleration diagram can be sketched. It yields the magnitude of $\boldsymbol{a}_B$:

$$a_B = a_A \tan\varphi + \frac{l\omega^2}{\cos\varphi}.$$

If we insert ω (which is known from the velocity diagram) we obtain the same result as in the analytical solution.

We will now show that the general *spatial motion* of a rigid body can always be decomposed into a translation and a rotation. To this end we introduce the $\bar{x}, \bar{y}, \bar{z}$-coordinate system as shown in Fig. 3.8. Its origin coincides with an arbitrary point A of the body and it undergoes a *translation* as the body moves, i.e., the directions of the axes remain unchanged (translating coordinate system). With respect to an observer located at the origin A of this system and fixed to the translating axes, the rigid body undergoes a rotation. The corresponding velocity and acceleration, respectively, of a point P are given by (3.5) and (3.6). In addition, we have to account for the velocity $\boldsymbol{v}_A$ and acceleration $\boldsymbol{a}_A$ of point A (i.e., of the $\bar{x}, \bar{y}, \bar{z}$-system) with respect to the *fixed* system x, y, z. Thus, the general motion of a rigid body in space is described by

$$
\begin{aligned}
\boldsymbol{r}_P &= \boldsymbol{r}_A + \boldsymbol{r}_{AP}, \\
\boldsymbol{v}_P &= \boldsymbol{v}_A + \boldsymbol{\omega} \times \boldsymbol{r}_{AP}, \\
\boldsymbol{a}_P &= \boldsymbol{a}_A + \dot{\boldsymbol{\omega}} \times \boldsymbol{r}_{AP} + \boldsymbol{\omega} \times (\boldsymbol{\omega} \times \boldsymbol{r}_{AP}).
\end{aligned}
\tag{3.8}
$$

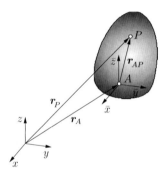

Fig. 3.8

Equations (3.8) are also valid in the case of a *plane* motion. Assuming that the motion takes place in the x, y-plane (cf. Fig. 3.5a), we can write

$$
\boldsymbol{\omega} = \omega \, \boldsymbol{e}_z, \quad \dot{\boldsymbol{\omega}} = \dot{\omega} \, \boldsymbol{e}_z, \quad \boldsymbol{r}_{AP} = r \, \boldsymbol{e}_r .
$$

Inserting into (3.8), we obtain Equations (3.7a,b) for the position, velocity and acceleration of P:

$$
\begin{aligned}
\boldsymbol{r}_P &= \boldsymbol{r}_A + r \, \boldsymbol{e}_r, \\
\boldsymbol{v}_P &= \boldsymbol{v}_A + \omega \, \boldsymbol{e}_z \times r \, \boldsymbol{e}_r = \boldsymbol{v}_A + r\omega \, \boldsymbol{e}_\varphi, \\
\boldsymbol{a}_P &= \boldsymbol{a}_A + \dot{\omega} \, \boldsymbol{e}_z \times r \, \boldsymbol{e}_r + \omega \, \boldsymbol{e}_z \times (\omega \, \boldsymbol{e}_z \times r \, \boldsymbol{e}_r) \\
&= \boldsymbol{a}_A + r\dot{\omega} \, \boldsymbol{e}_\varphi + \omega \, \boldsymbol{e}_z \times r\omega \, \boldsymbol{e}_\varphi = \boldsymbol{a}_A + r\dot{\omega} \, \boldsymbol{e}_\varphi - r\omega^2 \boldsymbol{e}_r .
\end{aligned}
$$

Example 3.1   A slider crank mechanism consists of a crankshaft $\overline{OA}$ and a connecting rod $\overline{AK}$ (Fig. 3.9a). The crankshaft rotates with a constant angular velocity ω_0.

Determine the angular velocity and the angular acceleration of the connecting rod as well as the velocity and the acceleration of the piston K in an arbitrary position.

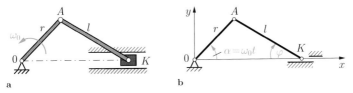

Fig. 3.9

Solution   We choose an x, y-coordinate system and the angles α and φ as shown in Fig. 3.9b. The piston K can move only in the horizontal direction. Therefore, its vertical displacement is zero:

$$y_K = r\sin\alpha - l\sin\varphi = 0 . \tag{a}$$

This yields the relation

$$\sin\varphi = \frac{r}{l}\sin\alpha \quad\rightarrow\quad \cos\varphi = \sqrt{1 - \frac{r^2}{l^2}\sin^2\alpha} \tag{b}$$

between the angles α and φ. If we differentiate (a), we obtain the angular velocity $\dot\varphi$ and the angular acceleration $\ddot\varphi$ of the connecting rod ($\dot\alpha = \omega_0 = \text{const}$):

$$\dot y_K = r\omega_0\cos\alpha - l\dot\varphi\cos\varphi = 0$$

$$\rightarrow\quad \underline{\underline{\dot\varphi = \omega_0\frac{r\cos\alpha}{l\cos\varphi}}},$$

$$\ddot y_K = -r\omega_0^2\sin\alpha + l\dot\varphi^2\sin\varphi - l\ddot\varphi\cos\varphi = 0$$

$$\rightarrow\quad \underline{\underline{\ddot\varphi}} = -\omega_0^2\frac{r\sin\alpha}{l\cos\varphi} + \dot\varphi^2\frac{\sin\varphi}{\cos\varphi}$$

$$= \omega_0^2\frac{r}{l}\left[-\frac{\sin\alpha}{\cos\varphi} + \frac{r\cos^2\alpha\sin\varphi}{l\ \cos^3\varphi}\right] .$$

The velocity $\dot{x}_K$ and the acceleration $\ddot{x}_K$ of the piston follow from the position x_K and (b):

$$x_K = r\cos\alpha + l\cos\varphi,$$

$$\underline{\underline{\dot{x}_K}} = -r\omega_0\sin\alpha - l\dot{\varphi}\sin\varphi = -r\omega_0\sin\alpha\left[1 + \frac{r\,\cos\alpha}{l\,\cos\varphi}\right],$$

$$\underline{\underline{\ddot{x}_K}} = -r\omega_0^2\cos\alpha - l\dot{\varphi}^2\cos\varphi - l\ddot{\varphi}\sin\varphi$$

$$= -r\omega_0^2\left[\cos\alpha - \frac{r}{l}\left(\frac{\sin^2\alpha}{\cos\varphi} - \frac{\cos^2\alpha}{\cos^3\varphi}\right)\right].$$

The angle φ can be replaced by the angle α with the aid of (b). If the velocity and the acceleration have to be determined as functions of the time t (instead of functions of the angle α), α has to be replaced by $\alpha = \omega_0 t$, where the initial condition $\alpha = 0$ for $t = 0$ has been assumed.

E3.2   **Example 3.2** The arm $\overline{0A}$ and the disk in Fig. 3.10a rotate with constant angular velocities ω_1 and ω_2, respectively.

Determine the velocity and acceleration of point P as functions of the angle ψ.

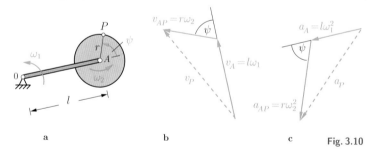

a   b   c   Fig. 3.10

Solution We solve the problem with a graphical/analytical method. The center A of the disk moves in a circle about 0. Thus,

$$v_A = l\omega_1 \quad \text{(perpendicular to } \overline{0A}\text{)},$$

$$a_A = l\omega_1^2 \quad \text{(in the direction from } A \text{ to } 0 \text{ since } \dot{\omega}_1 = 0\text{)}.$$

Point P follows a circular path about A:

$v_{AP} = r\omega_2$ (perpendicular to $\overline{AP}$),

$a_{AP} = r\omega_2^2$ (in the direction from P to A since $\dot{\omega}_2 = 0$).

Now we are able to sketch the velocity diagram (Fig. 3.10b) and the acceleration diagram (Fig. 3.10c). The law of cosines yields

$$v_P^2 = (l\omega_1)^2 + (r\omega_2)^2 - 2\,lr\omega_1\,\omega_2\cos(\pi - \psi)$$

$$\rightarrow \quad v_P = \sqrt{(l\omega_1)^2 + (r\omega_2)^2 + 2\,lr\omega_1\,\omega_2\cos\psi}$$

and

$$a_P = \sqrt{(l\omega_1^2)^2 + (r\omega_2^2)^2 + 2\,lr\omega_1^2\,\omega_2^2\cos\psi}\,.$$

Maximum (minimum) values are obtained for $\psi = 0$ ($\psi = \pi$). For example, the maximum acceleration is given by

$$a_{P_{\max}} = l\omega_1^2 + r\omega_2^2\,.$$

3.1.4 Instantaneous Center of Rotation

According to Section 3.1.3 plane motion of a rigid body is composed of a translation and a rotation. Alternatively, a plane motion may also be considered at each instant to be a pure rotation about a certain point Π. This point is referred to as the *instantaneous center of rotation* or the *instantaneous center of zero velocity*.

We will verify this statement by showing that there always exists a point A (= instantaneous center of rotation Π) which has a vanishing velocity. With $\boldsymbol{v}_A = \boldsymbol{0}$, (3.8) leads to the velocity (3.5) of an arbitrary point P during a pure rotation about A:

$$\boldsymbol{v}_P = \boldsymbol{\omega} \times \boldsymbol{r}_{AP}\,.$$

We can solve this equation for $\boldsymbol{r}_{AP}$ if we take the cross product of both sides with $\boldsymbol{\omega}$ and insert $\boldsymbol{v}_P = v_P\,\boldsymbol{e}_\varphi$, $\boldsymbol{r}_{AP} = r_P\,\boldsymbol{e}_r$ and

$\boldsymbol{\omega} = \omega \, \boldsymbol{e}_z$ ($\boldsymbol{e}_z$ is perpendicular to $\boldsymbol{e}_r$ and $\boldsymbol{e}_\varphi$):

$$\boldsymbol{0} = \boldsymbol{\omega} \times (\boldsymbol{\omega} \times \boldsymbol{r}_{AP}) - \boldsymbol{\omega} \times \boldsymbol{v}_P$$
$$= \omega^2 \, r_P \, \boldsymbol{e}_z \times (\boldsymbol{e}_z \times \boldsymbol{e}_r) - \omega \, v_P (\boldsymbol{e}_z \times \boldsymbol{e}_\varphi)$$
$$= -\omega^2 \, r_P \, \boldsymbol{e}_r - \omega \, v_P (\boldsymbol{e}_z \times \boldsymbol{e}_\varphi)$$
$$\rightarrow \quad \boldsymbol{r}_{AP} = r_P \, \boldsymbol{e}_r = -\frac{v_P}{\omega}(\boldsymbol{e}_z \times \boldsymbol{e}_\varphi) .$$

Thus, the vector $\boldsymbol{r}_{AP}$ is orthogonal to the velocity $\boldsymbol{v}_P$ (Fig. 3.11); it has the magnitude $r_P = v_P/\omega$. This uniquely determines the location of the instantaneous center of rotation Π. The instantaneous motion of a rigid body may therefore indeed be considered as being a pure rotation about Π.

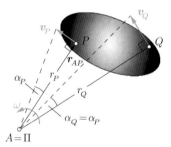

Fig. 3.11

The velocities of two arbitrary points P and Q can thus be given by (circular motion)

$$v_P = r_P \, \omega, \quad v_Q = r_Q \, \omega , \tag{3.9}$$

where ω is the magnitude of the angular velocity vector (which is orthogonal to the plane of the motion) and r_P and r_Q are the distances of P and Q from Π. The velocity vectors are perpendicular to the vectors $\boldsymbol{r}_{\Pi P}$ (from Π to P) and $\boldsymbol{r}_{\Pi Q}$ (from Π to Q), respectively (Fig. 3.11). Hence, the location of Π can be found if the directions of the velocities of two points of the rigid body are known: it is given by the point of intersection of the two straight lines which are perpendicular to the velocities. Note that the instantaneous center of rotation may lie outside the body. If there exists a point of the body with zero velocity, then this point is the instantaneous center of rotation. It should be emphasized that the instantaneous center of rotation is *not* a fixed point: it moves.

One can also eliminate the angular velocity ω from Equations (3.9). Then one obtains $v_P/r_P = v_Q/r_Q$ which means that the angles α_P and α_Q are equal (Fig. 3.11). This fact is used when problems are solved graphically with the aid of the instantaneous center of rotation (see Example 3.3).

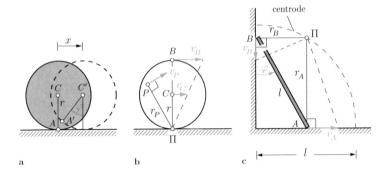

Fig. 3.12

As an illustrative example let us consider a wheel that moves on a horizontal plane. Its center C has the velocity v_C (Figs. 3.12a,b). We assume that the wheel *rolls* (no slipping at point A). As the wheel moves, its center C travels a distance x along a straight horizontal line, point A moves to the location A', and the wheel undergoes a rotation of angle φ. The arc length $r\varphi$ and the distance x have to coincide: $x = r\varphi$. Differentiating and using $\dot{x} = v_C$ and $\dot{\varphi} = \omega$ yields

$$v_C = r\,\omega. \tag{3.10}$$

The point of contact A with the ground momentarily has zero velocity (no slipping!): it is the instantaneous center of rotation Π (Fig. 3.12b). According to (3.9) and (3.10), the velocity of an arbitrary point P of the wheel is given by

$$v_P = r_P\, \omega = v_C \frac{r_P}{r}\,.$$

The velocity vector is perpendicular to the straight line $\overline{\Pi P}$. The maximum velocity is found at point B. With $r_B = 2\,r$ we get $v_B = 2\,v_C$.

Let us now reconsider the motion of the rod in Fig. 3.7a. The given velocity v_A is horizontal, the unknown velocity v_B is vertical (see Fig. 3.12c). The rod momentarily rotates about the instantaneous center of rotation Π which is given by the point of intersection of the straight lines which are perpendicular to the velocities. The angular velocity ω of the rod can immediately be obtained from (3.9):

$$v_A = r_A\, \omega = l\omega \cos\varphi \quad\rightarrow\quad \omega = \frac{v_A}{l \cos\varphi}\,.$$

This leads to the velocity of point B:

$$v_B = r_B\, \omega = l\omega \sin\varphi = v_A \tan\varphi\,.$$

As mentioned before, the instantaneous center of rotation Π is not a fixed point. Its location depends on the location of the rod. The locus of points which represent the instantaneous center of rotation in the space-fixed plane is called the *centrode*. In the present example it is a quarter-circle with radius l (Fig. 3.12c).

E3.3

Example 3.3 The pulley system shown in Fig. 3.13a consists of two pulleys ① and ② and a disk ③. The pulleys rotate with angular velocities ω_1 and ω_2, respectively.

Determine the angular velocity of the disk and the velocity of its center C. Assume that the disk does not slip on the cable.

Solution The velocities of the points A' and B' of the pulleys (Fig. 3.13b) are given by

$$v_{A'} = r_1\, \omega_1\,, \quad v_{B'} = r_2\, \omega_2\,.$$

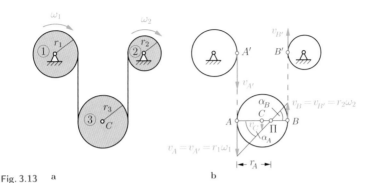

Fig. 3.13   a   b

Since there is no slipping of the cable, the velocities of the points A and B on the disk coincide with the velocities of A' and B':

$$v_A = v_{A'}, \quad v_B = v_{B'}.$$

To find the center of zero velocity Π of the disk, we first draw a straight line that is perpendicular to the direction of the velocities v_A and v_B. Then we connect the tips of the arrows of v_A and v_B with another straight line to ensure that the angles α_A and α_B are equal: $\alpha_A = \alpha_B$. The point of intersection of these straight lines is point Π (Fig. 3.13b). From the figure we obtain with the theorem of intersecting lines

$$\frac{r_A}{r_1\,\omega_1} = \frac{2\,r_3 - r_A}{r_2\,\omega_2} \quad \rightarrow \quad r_A = 2\,r_3\,\frac{r_1\,\omega_1}{r_1\,\omega_1 + r_2\,\omega_2}.$$

The angular velocity ω_3 of the disk follows from (3.9):

$$v_A = r_A\,\omega_3 \quad \rightarrow \quad \underline{\underline{\omega_3 = \frac{v_A}{r_A} = \frac{r_1\,\omega_1 + r_2\,\omega_2}{2\,r_3}}}.$$

The velocity of point C is given by

$$\underline{\underline{v_C}} = (r_A - r_3)\omega_3 = \frac{1}{2}(r_1\,\omega_1 - r_2\,\omega_2).$$

In the special case of $r_1\,\omega_1 = r_2\,\omega_2$ the motion of the disk is a pure rotation ($v_C = 0$).

E3.4

Example 3.4 Link $\textcircled{1}$ of the mechanism in Fig. 3.14a rotates with angular velocity ω_1.

Find the velocities of points A and B and the angular velocities of links $\textcircled{2}$ and $\textcircled{3}$ at the instant shown.

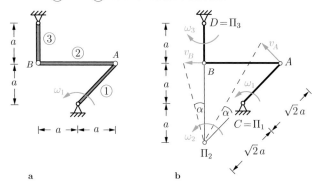

a   **b**

Fig. 3.14

Solution Links $\textcircled{1}$ and $\textcircled{3}$ rotate about the points C and D, respectively (Fig. 3.14b). Thus, these points are the centers Π_1 and Π_3 of zero velocity of the respective links. Therefore, the directions of the velocities of A and B are known. The points A and B are also points of link $\textcircled{2}$. Its center Π_2 of zero velocity is given by the point of intersection of the two straight lines that are perpendicular to v_A and v_B. From Fig. 3.14b we obtain

$$\Pi_1: \quad \underline{\underline{v_A = \sqrt{2}\,a\omega_1}},$$

$$\Pi_2: \quad v_A = 2\sqrt{2}\,a\omega_2 \quad \rightarrow \quad \underline{\underline{\omega_2 = \frac{\omega_1}{2}}},$$

$$v_B = 2\,a\omega_2 \quad \rightarrow \quad \underline{\underline{v_B = a\omega_1}},$$

$$\Pi_3: \quad v_B = a\omega_3 \quad \rightarrow \quad \underline{\underline{\omega_3 = \omega_1}}.$$

3.2 Kinetics of the Rotation about a Fixed Axis

In Section 3.1 we studied the motion of a rigid body without referring to forces as a cause or as a result of the motion. Now we will investigate motions under the influence of forces. In this section we restrict ourselves to the rotation about a fixed axis.

3.2.1 Principle of Angular Momentum

As the body in Fig. 3.15 rotates about the fixed axis a-a, each point of the body undergoes a circular motion. Therefore, the principle of angular momentum (1.67) for an infinitesimal mass element $\mathrm{d}m$ of the body is given by

$$\mathrm{d}\Theta_a\,\dot\omega = \mathrm{d}M_a \tag{3.11}$$

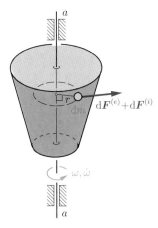

Fig. 3.15

where $\dot\omega = \ddot\varphi$. Here, $\mathrm{d}\Theta_a = r^2\mathrm{d}m$ is the *mass moment of inertia* of $\mathrm{d}m$ and $\mathrm{d}M_a$ is the sum of the moments of the external forces $\mathrm{d}\boldsymbol{F}^{(e)}$ and the internal forces $\mathrm{d}\boldsymbol{F}^{(i)}$ with respect to the axis of rotation. The superscripts in (1.67) that refer to the reference point are now replaced by subscripts which refer to the axis a-a. If we integrate over the complete body and assume that the moments of the internal forces cancel (cf. Section 2.3), then we obtain the *principle of angular momentum*

$$\Theta_a \, \dot{\omega} = M_a \, , \tag{3.12}$$

where

$$\Theta_a = \int r^2 \mathrm{d}m \tag{3.13}$$

is the mass moment of inertia of the body and M_a is the resultant moment of the external forces with respect to the axis a-a.

The angular momentum of a mass element $\mathrm{d}m$ is given by $\mathrm{d}L_a = rv \, \mathrm{d}m = r^2 \omega \, \mathrm{d}m$ (cf. Section 1.2.6). Integration yields the angular momentum of a rotating body with respect to the axis a-a:

$$L_a = \int \mathrm{d}L_a = \omega \int r^2 \mathrm{d}m \quad \rightarrow \quad L_a = \Theta_a \, \omega \, . \tag{3.14}$$

Thus, (3.12) can be written in the form

$$\dot{L}_a = M_a \, . \tag{3.15}$$

Integration from time t_0 to time t results in

$$L_a(t) - L_a(t_0) = \int_{t_0}^{t} M_a \, \mathrm{d}\bar{t} \quad \rightarrow \quad \Theta_a(\omega - \omega_0) = \int_{t_0}^{t} M_a \, \mathrm{d}\bar{t}. \tag{3.16}$$

Thus, the change of the angular momentum is equal to the time integral of the applied moment. In the special case of a vanishing moment M_a, the angular momentum $L_a = \Theta_a \, \omega$ does not change (conservation of angular momentum).

Equations (3.12), (3.15) and (3.16) are analogous to the equations of motion (1.38), (1.37) and (1.49) for a particle or for the translation of a rigid body. To obtain the equations for the rotation of a rigid body about a fixed axis, we have only to replace the mass with the mass moment of inertia, the velocity with the angular velocity, the force with the moment, and the linear momentum with the angular momentum. This is called an *analogy* between a translation and a rotation for fixed axis rotations (cf. Section 3.2.3).

3.2.2 Mass Moment of Inertia

According to (3.13) the mass moment of inertia, which is also called the *axial moment of inertia*, is defined as

$$\Theta_a = \int r^2 dm, \qquad (3.17)$$

where r is the perpendicular distance of an arbitrary element dm from the axis a-a.

In some cases it is helpful to use the *radius of gyration* r_g:

$$\Theta_a = r_g^2 \, m. \qquad (3.18)$$

One may interpret r_g as the distance from the axis a-a at which the total mass m can be imagined as being concentrated so that it has the same moment of inertia as the body.

If the density ρ of the material of the body is constant, then we use $dm = \rho \, dV$ and obtain from (3.17)

$$\Theta_a = \rho \int r^2 \, dV. \qquad (3.19a)$$

If, in addition, the shape of the cross section does not change through the length of the body (example: cylinder, see Fig. 3.16a), then we get with $dV = l \, dA$

$$\Theta_a = \rho l \int r^2 \, dA = \rho l \, I_p. \qquad (3.19b)$$

Here, I_p is the *polar moment of area* (see Volume 2, Section 4.2).

The mass moments of inertia obey the *parallel-axis theorem*, just as the second moments of area do (see Volume 2, Section 4.2.2). In order to derive this theorem, we consider two parallel axes as shown in Fig. 3.16b. The axis c-c passes through the center of mass C of the body. With $x = x_c + \bar{x}$ and $y = y_c + \bar{y}$ (Fig. 3.16c) we obtain

$$\Theta_a = \int r^2 dm = \int (x^2 + y^2) \, dm = (x_c^2 + y_c^2) \int dm$$
$$+ 2 x_c \int \bar{x} \, dm + 2 y_c \int \bar{y} \, dm + \int (\bar{x}^2 + \bar{y}^2) \, dm.$$

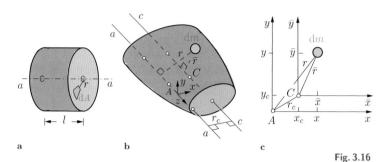

Fig. 3.16

The first moments $\int \bar{x}\,\mathrm{d}m$ and $\int \bar{y}\,\mathrm{d}m$ about axes through the center of mass are zero (cf. Volume 1, Section 4.3). Therefore, using

$$\Theta_c = \int \bar{r}^2\mathrm{d}m = \int (\bar{x}^2 + \bar{y}^2)\,\mathrm{d}m, \quad x_c^2 + y_c^2 = r_c^2, \quad m = \int \mathrm{d}m$$

the parallel-axis theorem is obtained:

$$\Theta_a = \Theta_c + r_c^2\, m\,. \tag{3.20}$$

From (3.18) we find the relation $r_{ga}^2 = r_{gc}^2 + r_c^2$ for the radii of gyration.

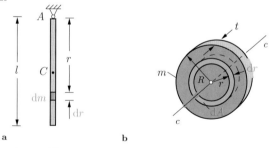

Fig. 3.17

As an illustrative example we consider a slender homogeneous rod (mass m), see Fig. 3.17a. With $\mathrm{d}m/m = \mathrm{d}r/l$ we obtain the moment of inertia

$$\Theta_A = \int r^2\,\mathrm{d}m = \frac{m}{l}\int_0^l r^2\,\mathrm{d}r = \frac{ml^2}{3} \tag{3.21a}$$

with respect to an axis that is perpendicular to the rod and passes through point A. Here, we replaced the subscript a (reference axis) by the subscript A (reference point). In the following we will use both notations. If we choose a reference axis that passes through the center C of the rod, (3.20) leads to

$$\Theta_C = \Theta_A - \left(\frac{l}{2}\right)^2 m = \frac{ml^2}{12} . \tag{3.21b}$$

We now determine the moment of inertia of a homogeneous circular disk with mass m, thickness t, and radius R. We choose the reference axis c-c that is perpendicular to the plane of the disk and passes through its center (Fig. 3.17b). With the area element $\mathrm{d}A = 2\,\pi r\,\mathrm{d}r$ we obtain

$$\Theta_c = \int r^2\,\mathrm{d}m = \rho t \int r^2\,\mathrm{d}A = 2\,\pi\rho t \int_0^R r^3\,\mathrm{d}r = \frac{\pi}{2}\rho t R^4 = \frac{mR^2}{2} .$$

$$\tag{3.22}$$

The moment of inertia Θ_c depends on the mass m and the radius R; it is independent of the thickness t. Therefore, the result (3.22) is also valid for a circular cylinder of arbitrary length.

Example 3.5 Determine the moment of inertia of a homogeneous solid sphere (mass m, radius R) with respect to an axis that passes through the center C.

E 3.5

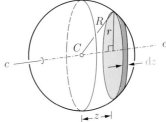

Fig. 3.18

Solution We consider the sphere as being composed of circular disks with infinitesimal thickness $\mathrm{d}z$ (Fig. 3.18). According to (3.22), the moment of inertia of a disk (radius $r = \sqrt{R^2 - z^2}$) with respect to the axis c-c is given by

$$\mathrm{d}\Theta_c = \frac{1}{2}\mathrm{d}m\,r^2 = \frac{1}{2}(\rho r^2\,\pi\,\mathrm{d}z)\,r^2 = \frac{\pi}{2}\,\rho(R^2 - z^2)^2\,\mathrm{d}z .$$

Integration yields the moment of inertia of the sphere:

$$\Theta_c = \int \mathrm{d}\Theta_c = \frac{\pi}{2}\rho \int\limits_{-R}^{+R} (R^2 - z^2)^2 \, \mathrm{d}z = \frac{8}{15}\pi\rho R^5 \, .$$

With the mass $m = \rho V$ and the volume $V = \frac{4}{3}\pi R^3$ of the sphere, this can be written as

$$\Theta_c = \frac{2}{5}\, m\, R^2 \, .$$

E3.6   **Example 3.6** A homogeneous square plate (weight $W = mg$) is suspended by a pin at A (Fig. 3.19a). It is displaced from the position of equilibrium and then released.
    Find the equation of motion.

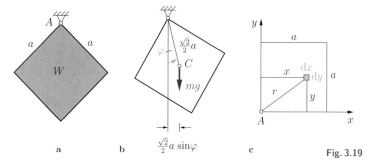

a   **b**   **c**   Fig. 3.19

Solution The plate rotates about the fixed axis that is perpendicular to the plane of the plate and passes through A. We introduce the angle φ as shown in Fig. 3.19b and apply the principle of angular momentum:

$$\curvearrowleft A: \quad \Theta_A \ddot{\varphi} = M_A \tag{a}$$

with

$$M_A = -mg\frac{\sqrt{2}}{2}a\sin\varphi \, . \tag{b}$$

The symbol $\curvearrowleft A$: characterizes the point of reference and the positive sense of rotation.

    To find the moment of inertia we use (3.17) and the notation in Fig. 3.19c. With the thickness t of the plate and

$$dm = \rho \, dV = \rho t \, dx \, dy, \quad m = \rho a^2 t, \quad r^2 = x^2 + y^2$$

we obtain

$$\Theta_A = \int r^2 \, dm = \rho t \int\limits_0^a \int\limits_0^a (x^2 + y^2) \, dx \, dy = \frac{2}{3} \rho t a^4 = \frac{2}{3} m a^2 .$$

Inserting Θ_A and M_A (see (b)) into (a), the equation of motion becomes

$$\frac{2}{3} m a^2 \, \ddot{\varphi} = - mg \frac{\sqrt{2}}{2} a \sin \varphi \quad \rightarrow \quad \underline{\underline{\ddot{\varphi} + \frac{3\sqrt{2}}{4} \frac{g}{a} \sin \varphi = 0}} .$$

3.2.3 Work, Energy, Power

Assume that a rigid body rotates about a fixed axis a-a. Then the kinetic energy T of the body follows from (1.69) with $v = r\omega$:

$$T = \frac{1}{2} \int v^2 \, dm = \frac{1}{2} \omega^2 \int r^2 \, dm$$

or

$$T = \frac{1}{2} \Theta_a \, \omega^2 . \tag{3.23}$$

The moment M_a of the external forces does work $dU = M_a d\varphi$ during an infinitesimal rotation of angle $d\varphi$. Therefore, the work during a finite rotation from φ_0 to φ is obtained as the integral

$$U = \int\limits_{\varphi_0}^{\varphi} M_a d\bar{\varphi} \tag{3.24}$$

and the power is given by

$$P = \frac{dU}{dt} = M_a \, \omega . \tag{3.25}$$

If we integrate the principle of angular momentum (3.12) from φ_0 to φ, we get the work-energy theorem (note that $d\varphi = \omega dt$ and $\dot{\omega}\omega = (\frac{1}{2}\omega^2)^{\cdot}$)

$$\Theta_a \int_{\varphi_0}^{\varphi} \dot{\omega} d\bar{\varphi} = \int_{\varphi_0}^{\varphi} M_a d\bar{\varphi} \rightarrow \Theta_a \int_{t_0}^{t} \dot{\omega}\omega d\bar{t} = \frac{1}{2}\Theta_a \omega^2 - \frac{1}{2}\Theta_a \omega_0^2 = \int_{\varphi_0}^{\varphi} M_a d\bar{\varphi}$$

or

$$T - T_0 = U. \tag{3.26}$$

In the special case that the moment M_a can be derived from a potential V, we have $U = -(V - V_0)$ and obtain a statement of conservation of energy:

$$T + V = T_0 + V_0 = \text{const}. \tag{3.27}$$

Table 3.1

Translation		Rotation about a fixed axis a-a	
s	displacement	angle	φ
$v = \dot{s}$	velocity	angular velocity	$\omega = \dot{\varphi}$
$a = \dot{v} = \ddot{s}$	acceleration	angular acceleration	$\dot{\omega} = \ddot{\varphi}$
m	mass	moment of inertia	Θ_a
F	force (in the direction of the displacement)	moment (about a-a)	M_a
$p = mv$	linear momentum	angular momentum	$L_a = \Theta_a \omega$
$ma = F$	principle of linear momentum	principle of angular momentum	$\Theta_a \dot{\omega} = M_a$
$T = \frac{1}{2}mv^2$	kinetic energy		$T = \frac{1}{2}\Theta_a \omega^2$
$W = \int F \, ds$	work		$U = \int M_a \, d\varphi$
$P = Fv$	power		$P = M_a \omega$

In Section 3.2.1 it was already mentioned that there exists an analogy between the rotation of a rigid body about a fixed axis and the translation of a particle (a body). Accordingly, we obtain the equations for the rotation from those of the translation if we replace the mass by the moment of inertia, the velocity by the angular velocity, the force by the moment, etc. This is also valid for the quantities that were derived in this section (work, energy, power) and for the principles (e.g., the conservation of energy). Table 3.1 shows these analogies.

Example 3.7 A drum (moment of inertia Θ_A) rotates at time t_0 with the angular velocity ω_0. For $t > t_0$, a brake (coefficient of kinetic friction μ) acts on the drum (Fig. 3.20a).

E3.7

Determine the number of rotations until the drum comes to a stop. Assume $F = \text{const}$.

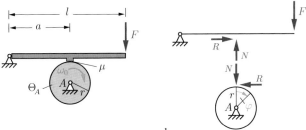

Fig. 3.20   a   b

Solution We separate the drum and the lever; the free-body diagram is shown in Fig. 3.20b (note: R is the friction force). Moment equilibrium of the lever yields the normal force

$$N = \frac{l}{a} F.$$

If we apply the principle of angular momentum (3.12) to the drum, we obtain

$$\overset{\curvearrowright}{A}: \quad \Theta_A \ddot{\varphi} = -rR.$$

Using Coulomb's law of friction $R = \mu N = \mu \dfrac{l}{a} F$ leads to

$$\ddot{\varphi} = -\kappa,$$

where the parameter $\kappa = \dfrac{r\mu l F}{a\Theta_A}$ has been introduced. We use the initial conditions $\dot{\varphi}(0) = \omega_0$, $\varphi(0) = 0$ and integrate twice:

$$\dot{\varphi} = -\kappa t + \omega_0, \quad \varphi = -\frac{1}{2}\kappa t^2 + \omega_0 t .$$

The time t_s, the angle φ_s and the number of rotations n_s until the drum comes to a stop follow from the condition $\dot{\varphi} = 0$:

$$t_s = \frac{\omega_0}{\kappa}, \quad \varphi_s = \varphi(t_s) = \frac{\omega_0^2}{2\kappa}, \quad n_s = \frac{\varphi_s}{2\pi} = \underline{\underline{\frac{\omega_0^2}{4\pi\kappa}}} .$$

The problem can also be solved with the aid of the work-energy theorem. The kinetic energy of the drum at time $t_0 = 0$ (i.e., $\varphi = 0$) is given by

$$T_0 = \frac{1}{2}\Theta_A \omega_0^2 .$$

At time t_s (i.e., $\varphi = \varphi_s$) we have

$$T_s = 0.$$

The work of the external forces done during the time interval from 0 to t_s is

$$U = \int_0^{\varphi_s} M_A \,\mathrm{d}\varphi = -\int_0^{\varphi_s} r\,R\,\mathrm{d}\varphi = -r\,R\,\varphi_s.$$

Thus, (3.26) yields

$$-\frac{1}{2}\Theta_A \omega_0^2 = -r\,R\,\varphi_s \quad \rightarrow \quad \underline{\underline{n_s}} = \frac{\varphi_s}{2\pi} = \frac{\Theta_A \omega_0^2}{4\pi r R} = \underline{\underline{\frac{\omega_0^2}{4\pi\kappa}}} .$$

E3.8

Example 3.8 Determine the velocity of the block (mass m_1) as a function of its displacement if the system in Fig. 3.21a is released from rest. Neglect the mass of the pulley R and assume both pulleys are ideal (frictionless).

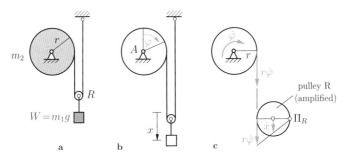

Fig. 3.21

Solution   The only external force (the weight W) acting on the system is a conservative force. Since the velocity has to be determined as a function of the displacement of the block, it is of advantage to use the conservation of energy:

$$T + V = T_0 + V_0\,.$$

We denote the displacement of the block from the initial position with x and the corresponding rotation of the upper pulley with φ (Fig. 3.21b). Then the kinetic and the potential energy are given by

$$T_0 = 0, \quad V_0 = 0$$

in the initial position and by

$$T = \frac{1}{2}\, m_1\, \dot{x}^2 + \frac{1}{2}\, \Theta_A\, \dot{\varphi}^2, \quad V = -\, m_1\, g x$$

in a displaced position. The kinetic energy results from the translation of the block and the rotation of the pulley. With the kinematic relation (see Fig. 3.21c)

$$\dot{x} = \frac{1}{2}\, r\, \dot{\varphi} \quad \rightarrow \quad \dot{\varphi} = 2\frac{\dot{x}}{r}$$

and the moment of inertia $\Theta_A = \frac{1}{2} m_2 r^2$ (see (3.22)) we find

$$\left[\frac{1}{2}\, m_1\, \dot{x}^2 + \frac{1}{2}\left(\frac{1}{2} m_2\, r^2 \right)\left(4\,\frac{\dot{x}^2}{r^2} \right) \right] - m_1\, g x = 0$$

$$\rightarrow \quad v = \dot{x} = \pm \sqrt{\frac{2\,m_1}{m_1 + 2\,m_2}\,gx}\,.$$

In the special case of a negligible mass of the pulley ($m_2 \ll m_1$) we obtain the velocity $v_2 = \sqrt{2\,gx}$ of the free fall of a point mass.

3.3 Kinetics of a Rigid Body in Plane Motion

3.3.1 Principles of Linear and Angular Momentum

Let us consider a rigid body whose particles move in the x, y-plane or in a plane parallel to it (Fig. 3.22). Then the body is said to undergo a plane (planar) motion. The external force $d\boldsymbol{F}$ that acts on a mass element dm has the components dF_x and dF_y. Internal forces need not be taken into account in the case of a rigid body (cf. Chapter 2). Let A be an arbitrary point that is fixed in the body. Since

$$\xi = r \cos \varphi, \quad \eta = r \sin \varphi \tag{3.28}$$

(see Fig. 3.22) the coordinates of the element dm are given by

$$x = x_A + \xi = x_A + r \cos \varphi, \quad y = y_A + \eta = y_A + r \sin \varphi\,.$$

If we differentiate these equations, use $\dot{\varphi} = \omega$ and (3.28), we obtain the components of the velocity and acceleration:

$$\dot{x} = \dot{x}_A - r\omega \sin \varphi = \dot{x}_A - \omega\eta, \qquad \dot{y} = \dot{y}_A + r\omega \cos \varphi = \dot{y}_A + \omega\xi, \tag{3.29a}$$

$$\ddot{x} = \ddot{x}_A - r\dot{\omega} \sin \varphi - r\omega^2 \cos \varphi \qquad \ddot{y} = \ddot{y}_A + r\dot{\omega} \cos \varphi - r\omega^2 \sin \varphi$$

$$= \ddot{x}_A - \dot{\omega}\eta - \omega^2\xi, \qquad \qquad = \ddot{y}_A + \dot{\omega}\xi - \omega^2\eta\,. \tag{3.29b}$$

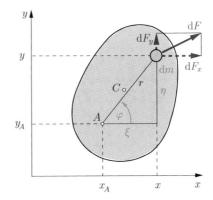

Fig. 3.22

Thus, the equations of motion for the element $\mathrm{d}m$ are

$$\ddot{x}\,\mathrm{d}m = \ddot{x}_A\,\mathrm{d}m - \dot{\omega}\,\eta\,\mathrm{d}m - \omega^2\,\xi\,\mathrm{d}m = \mathrm{d}F_x,$$
$$\ddot{y}\,\mathrm{d}m = \ddot{y}_A\,\mathrm{d}m + \dot{\omega}\,\xi\,\mathrm{d}m - \omega^2\,\eta\,\mathrm{d}m = \mathrm{d}F_y\,.$$

Integration yields the forces F_x, F_y and the moment M_A with respect to point A (note the positive sense of rotation!):

$$F_x = \int \mathrm{d}\,F_x = \ddot{x}_A \int \mathrm{d}m - \dot{\omega} \int \eta\,\mathrm{d}m - \omega^2 \int \xi\,\mathrm{d}m,$$
$$F_y = \int \mathrm{d}\,F_y = \ddot{y}_A \int \mathrm{d}m + \dot{\omega} \int \xi\,\mathrm{d}m - \omega^2 \int \eta\,\mathrm{d}m, \tag{3.30a}$$

$$M_A = \int \xi\,\mathrm{d}F_y - \int \eta\,\mathrm{d}F_x = \ddot{y}_A \int \xi\,\mathrm{d}m + \dot{\omega} \int \xi^2\,\mathrm{d}m$$
$$- \omega^2 \int \xi\eta\,\mathrm{d}m - \ddot{x}_A \int \eta\,\mathrm{d}m + \dot{\omega} \int \eta^2\,\mathrm{d}m + \omega^2 \int \xi\eta\,\mathrm{d}m\,. \tag{3.30b}$$

We now choose point A to be the center of mass C of the body. Then the first moments $\int \xi\,\mathrm{d}m$ and $\int \eta\,\mathrm{d}m$ are zero. With $m = \int \mathrm{d}m$ and $\Theta_C = \int r^2\,\mathrm{d}m = \int(\xi^2+\eta^2)\mathrm{d}m$ Equations (3.30a,b) reduce to

$$m\ddot{x}_c = F_x, \quad m\ddot{y}_c = F_y, \tag{3.31a}$$

$$\Theta_C\,\ddot{\varphi} = M_C\,. \tag{3.31b}$$

Here, F_x and F_y are the resultants in the x- and y-directions, respectively, of the external forces and M_C is the resultant external moment with respect to C. Equations (3.31a) describe the motion of the center of mass. They are analogous to the equations of motion (1.38) of a particle and are referred to as the *principle of linear momentum*. Equation (3.31b) describes the rotation of the body about the mass center; it represents the *principle of angular momentum* . The point of reference thereby has to be the center of mass C. The principles of linear and angular momentum describe the general plane motion of a rigid body. In the special case of a body at rest ($\ddot{x}_c = 0$, $\ddot{y}_c = 0$, $\ddot{\varphi} = 0$) Equations (3.31a,b) reduce to the equilibrium conditions in statics.

If the body undergoes a translation ($\dot{\varphi} = 0, \ddot{\varphi} = 0$), then (3.31b) requires

$$M_C = 0 \,. \tag{3.32a}$$

Hence, the moment of the external forces with respect to the center of mass C has to be zero. Then the motion of C and thus of any point of the body can be found from

$$m\ddot{x}_c = F_x, \quad m\ddot{y}_c = F_y \,. \tag{3.32b}$$

We now refer to the special case of a planar motion where the body undergoes a pure rotation about a fixed point A. Then we obtain from (3.30b) with $\ddot{x}_A = \ddot{y}_A = 0$ and $\int(\xi^2 + \eta^2)\mathrm{d}m = \Theta_A$

$$\Theta_A \, \ddot{\varphi} = M_A \,. \tag{3.33}$$

This is the result that we already found in the case of a rotation about a fixed axis (see Section 3.2.1). The axis is perpendicular to the x, y-plane and it passes through point A. In this case, the point of reference in the principle of angular momentum may either be the center of mass C or the fixed point A.

As an illustrative example we consider a homogeneous solid sphere that moves down a rough inclined plane (Fig. 3.23a). We first assume that the sphere rolls without slipping. The free-body diagram (Fig. 3.23b) shows the forces acting on the sphere. The coordinates that describe the motion are the position x_c of the center of the sphere and the angle φ of rotation. With $\ddot{y}_c = 0$,

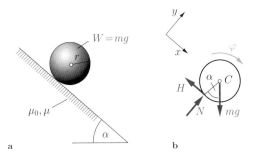

Fig. 3.23

the principles of linear and angular momentum yield

$$\searrow: \quad m\ddot{x}_c = mg\sin\alpha - H, \tag{a}$$

$$\nearrow: \quad 0 = N - mg\cos\alpha \quad \rightarrow \quad N = mg\cos\alpha, \tag{b}$$

$$\overset{\curvearrowright}{C}: \quad \Theta_C\,\ddot{\varphi} = r\,H. \tag{c}$$

The moment of inertia of the sphere is given by $\Theta_C = \frac{2}{5}mr^2$ (see Example 3.5).

Since it is assumed that the sphere rolls without slipping, the kinematic relation

$$\dot{x}_c = r\dot{\varphi} \quad \rightarrow \quad \ddot{\varphi} = \frac{\ddot{x}_c}{r} \tag{d}$$

holds (see (3.10)). Thus, (a), (c), and (d) yield the acceleration of the center of mass C:

$$m\ddot{x}_c = mg\sin\alpha - \frac{\Theta_C}{r^2}\ddot{x}_c \quad \rightarrow \quad \ddot{x}_c = \frac{g\sin\alpha}{1 + \dfrac{\Theta_C}{mr^2}} = \frac{5}{7}g\sin\alpha\,.$$

The friction force during rolling follows from (a):

$$H = m(g\sin\alpha - \ddot{x}_c) = \frac{2}{7}mg\sin\alpha\,.$$

Now we are able to formulate the condition which must be satisfied by the coefficient of static friction μ_0 in order to ensure rolling of the sphere:

$$H \leqq \mu_0 N \quad \rightarrow \quad \mu_0 \geqq \frac{H}{N} = \frac{\frac{2}{7} mg \sin \alpha}{mg \cos \alpha} = \frac{2}{7} \tan \alpha .$$

If μ_0 does not satisfy this condition, the sphere will *slip* on the inclined plane. Then the friction force H has to be replaced by the dynamic friction force R in Fig. 3.23b and in (a) and (c):

$$m\ddot{x}_c = mg \sin \alpha - R, \quad N = mg \cos \alpha, \quad \Theta_C \ddot{\varphi} = r R . \qquad \text{(e)}$$

In addition we have to use the friction law

$$R = \mu N . \qquad \text{(f)}$$

When the sphere is slipping, *no* relation exists between $\ddot{x}_c$ and $\ddot{\varphi}$: they are independent of each other. From (e) and (f) we obtain the accelerations

$$\ddot{x}_c = g(\sin \alpha - \mu \cos \alpha), \quad \ddot{\varphi} = \frac{5 \mu g}{2 r} \cos \alpha .$$

E3.9 **Example 3.9** A simplified model of a car is shown in Fig. 3.24a. It consists of a rigid body (weight $W = mg$, center of mass at C) and massless wheels.

Find the maximum acceleration of the car on a rough horizontal surface (coefficient of static friction μ_0), if the engine only drives a) the rear wheels (the front wheels are rolling freely), b) the front wheels (the rear wheels are rolling freely).

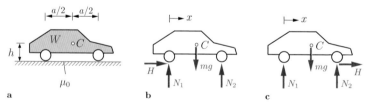

Fig. 3.24

Solution a) Fig. 3.24b shows the free-body diagram if the engine drives the rear wheels. Since the rigid body undergoes a translation in the *horizontal* direction, the forces in the *vertical* direction and the moments with respect to the center of mass have to be in equilibrium. Thus, the principles of linear and angular momentum yield

$$\rightarrow : \quad m\ddot{x} = H,$$

$$\uparrow : \quad 0 = N_1 + N_2 - mg,$$

$$\overset{\curvearrowright}{C} : \quad 0 = \frac{a}{2} N_1 - \frac{a}{2} N_2 - hH \,,$$

which leads to

$$N_1 = \frac{mg}{2} + \frac{h}{a} H, \quad N_2 = \frac{mg}{2} - \frac{h}{a} H \,.$$

We now consider the limiting case of impending slip at the rear wheels. Then the condition

$$H_{\max} = \mu_0 N_1$$

has to be satisfied. We insert N_1 to obtain

$$H_{\max} = \mu_0 \left[\frac{mg}{2} + \frac{h}{a} H_{\max} \right] \quad \rightarrow \quad H_{\max} = \frac{mg}{2} \frac{\mu_0}{1 - \mu_0 \dfrac{h}{a}} \,.$$

Since $m\ddot{x}_{\max} = H_{\max}$, the maximum acceleration is found to be

$$\underline{\underline{\ddot{x}_{\max} = \frac{g}{2} \frac{\mu_0}{1 - \mu_0 \dfrac{h}{a}}}} \,.$$

Note that this result is not valid if $N_2 < 0$. Then the front wheels lift off the ground.

b) We now assume that the engine drives the front wheels (Fig. 3.24 c). Then, the principles of linear and angular momentum remain unchanged, whereas the condition of limiting friction is now given by

$$H_{\max} = \mu_0 N_2 \,.$$

This yields the maximum acceleration

$$\ddot{x}_{\max} = \frac{g}{2}\frac{\mu_0}{1 + \mu_0\dfrac{h}{a}}$$

which, for this model, is smaller than the one obtained in part a) of the example.

E3.10   **Example 3.10** A wheel (weight $W = mg$, moment of inertia Θ_C) rolls without slipping on its hub on a horizontal track (Fig. 3.25a).

Determine the acceleration of the center of mass C and the contact forces between the wheel and the track. Assume $F = $ const.

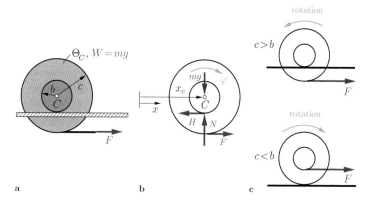

Fig. 3.25

Solution The free-body diagram (Fig. 3.25b) shows the forces that act on the wheel. We introduce the coordinates x and φ. Then the principles of linear and angular momentum yield

$$\rightarrow :\quad m\ddot{x}_c = F - H,$$

$$\uparrow :\quad 0 = N - mg,$$

$$\overset{\curvearrowright}{C} :\quad \Theta_C\,\ddot{\varphi} = bH - cF\,.$$

In addition, we have the kinematic relation

$$\dot{x}_c = b\,\dot{\varphi} \quad \rightarrow \quad \ddot{\varphi} = \frac{\ddot{x}_c}{b}$$

between $\dot{x}_c$ and $\dot{\varphi}$ (rolling without slip). Thus, we have four equations for the four unknowns $N, H, \ddot{x}_c$ and $\ddot{\varphi}$. We solve the equations to obtain the acceleration of the center of mass:

$$\underline{\underline{\ddot{x}_c = -\frac{F\left(\dfrac{c}{b} - 1\right)}{m\left(1 + \dfrac{\Theta_C}{mb^2}\right)}}}.$$

The acceleration is negative for $c > b$ (motion to the left) and it is positive for $c < b$ (motion to the right). The directions of the motion are illustrated in Fig. 3.25c.

The contact forces between the wheel and the track are found to be

$$\underline{\underline{N = mg}}, \quad \underline{\underline{H = F\,\frac{1 + \dfrac{mb^2}{\Theta_C}\dfrac{c}{b}}{1 + \dfrac{mb^2}{\Theta_C}}}}.$$

Example 3.11 A homogeneous slender rod (weight $W = mg$) is pin-supported at point A (Fig. 3.26a). It is released from rest when it is horizontal.

    Find the angular acceleration, the angular velocity and the support reactions as functions of the position of the rod.

E3.11

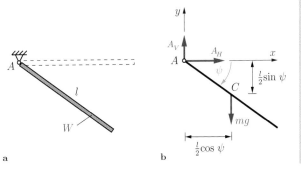

Fig. 3.26   a   b

Solution The rod rotates about the fixed point A. Therefore, we apply the principle of angular momentum (3.12). With the angle ψ as introduced in Fig. 3.26b we obtain

$$\widehat{A} : \quad \Theta_A \ddot{\psi} = mg \frac{l}{2} \cos \psi \,.$$

This yields the angular acceleration

$$\underline{\underline{\ddot{\psi} = \frac{3}{2} \frac{g}{l} \cos \psi}}$$

where the moment of inertia $\Theta_A = ml^2/3$ has been introduced.

The angular velocity is obtained through integration and application of $\ddot{\psi} = \dfrac{\mathrm{d}\dot{\psi}}{\mathrm{d}\psi} \dot{\psi}$ (compare Section 1.1.3):

$$\frac{\dot{\psi}^2}{2} = \int \ddot{\psi} \, \mathrm{d}\psi = \frac{3}{2} \frac{g}{l} \sin \psi + K \,.$$

To calculate the constant of integration K we use the initial condition $\dot{\psi}(\psi = 0) = 0$ which yields $K = 0$. Thus,

$$\underline{\underline{\dot{\psi} = \pm \sqrt{3 \frac{g}{l} \sin \psi}}} \,.$$

The support reactions follow from the principle of linear momentum:

$$\rightarrow: \quad m\ddot{x}_c = A_H \,, \qquad \uparrow: \quad m\ddot{y}_c = A_V - mg \,.$$

Starting with the coordinates of the center of mass C (Fig. 3.26b) we obtain the acceleration of C through differentiation and insertion of $\dot{\psi}$ and $\ddot{\psi}$:

$$x_c = \frac{l}{2} \cos \psi \,, \qquad\qquad y_c = -\frac{l}{2} \sin \psi \,,$$

$$\dot{x}_c = -\frac{l}{2} \dot{\psi} \sin \psi \,, \qquad\qquad \dot{y}_c = -\frac{l}{2} \dot{\psi} \cos \psi \,,$$

$$\ddot{x}_c = -\frac{l}{2} \ddot{\psi} \sin \psi - \frac{l}{2} \dot{\psi}^2 \cos \psi \qquad \ddot{y}_c = -\frac{l}{2} \ddot{\psi} \cos \psi + \frac{l}{2} \dot{\psi}^2 \sin \psi$$

$$= -\frac{9}{8} g \sin 2\psi \,, \qquad\qquad\qquad = \frac{3}{8} g (1 - 3 \cos 2\psi) \,.$$

Hence,

$$\underline{\underline{A_H}} = m\ddot{x}_c = -\frac{9}{8} W \sin 2\psi,$$

$$\underline{\underline{A_V}} = mg + m\ddot{y}_c = W \left(\frac{11}{8} - \frac{9}{8} \cos 2\psi\right).$$

In the initial position (horizontal rod: $\psi = 0$) we get $A_H = 0$ and $A_V = W/4$. For $\psi = \pi/2$ (vertical rod) we find $A_H = 0$ and $A_V = 5W/2$.

Example 3.12   A block (mass m_1) can move horizontally on a smooth surface (Fig. 3.27a). A homogeneous rod (mass m_2) is connected to the block by a pin. The rod is displaced from its equilibrium position and then released.

E3.12

Find the equations of motion for the special case $m_1 = m_2$.

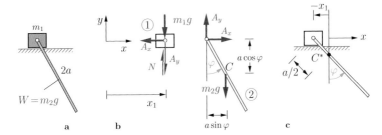

Fig. 3.27

Solution   The system has two degrees of freedom. We separate the two rigid bodies in the free-body diagram and we introduce a coordinate system x, y and the angle φ (Fig. 3.27b). For block ① we obtain

$$\rightarrow:\quad m_1\ddot{x}_1 = -A_x,$$

$$\uparrow:\quad 0 = N - m_1 g - A_y$$

(a)

and for rod ②

$$\rightarrow:\quad m_2\ddot{x}_c = A_x,$$

(b)

$$\uparrow: \quad m_2 \ddot{y}_c = A_y - m_2\, g, \tag{c}$$

$$\curvearrowleft \atop C: \quad \Theta_C \ddot{\varphi} = -a \cos\varphi\, A_x - a \sin\varphi\, A_y \tag{d}$$

where

$$\Theta_C = \frac{m_2(2\,a)^2}{12} = \frac{m_2\, a^2}{3}.$$

The motion of the block and the motion of the rod are kinematically related:

$$x_c = x_1 + a \sin\varphi, \qquad\qquad y_c = -a \cos\varphi,$$

$$\dot{x}_c = \dot{x}_1 + a\, \dot\varphi \cos\varphi, \qquad\qquad \dot{y}_c = a\dot\varphi \sin\varphi, \tag{e}$$

$$\ddot{x}_c = \ddot{x}_1 + a\, \ddot\varphi \cos\varphi - a\dot\varphi^2 \sin\varphi, \quad \ddot{y}_c = a\, \ddot\varphi \sin\varphi + a\dot\varphi^2 \cos\varphi.$$

We can solve this system of equations to find the equation of motion for the angle $\varphi(t)$. Using (e), (a) and $m_1 = m_2 = m$, we find from (b) and (c)

$$A_x = \frac{ma}{2}(\ddot\varphi \cos\varphi - \dot\varphi^2 \sin\varphi),$$
$$A_y = mg + ma(\ddot\varphi \sin\varphi + \dot\varphi^2 \cos\varphi). \tag{f}$$

Substituting into (d) leads to

$$\ddot\varphi(8 - 3\cos^2\varphi) + 3\dot\varphi^2 \sin\varphi \cos\varphi + 6\frac{g}{a}\sin\varphi = 0.$$

If a solution of this nonlinear differential equation is known, we can find the displacement x_1 from (f) and (a).

A kinematic relation between x_1 and φ may also be obtained in the following way. There is no external horizontal force acting on the system. Therefore the center of gravity C^* (Fig. 3.27c) of the complete system can not undergo a horizontal displacement (if it was at rest initially). If we now measure the coordinate x from C^* (at a distance $a/2$ from the pin), we obtain the relation $x_1 = -a/2 \sin\varphi$. In this example, the motion of the system with two degrees of freedom can be described by only φ or only x_1.

3.3.2 Impulse Laws, Work-Energy Theorem and Conservation of Energy

We will now integrate Equations (3.31a,b) from t_0 to t. Using the notations $x_{c0} = x_c(t_0)$ etc, we obtain

$$m\dot{x}_c - m\dot{x}_{c0} = \hat{F}_x, \quad m\dot{y}_c - m\dot{y}_{c0} = \hat{F}_y, \tag{3.34a}$$

$$\Theta_C \dot{\varphi} - \Theta_C \dot{\varphi}_0 = \hat{M}_C. \tag{3.34b}$$

Equations (3.34a,b) are referred to as the *Impulse Laws*. In particular, (3.34a) is the *principle of linear impulse and momentum*; (3.34b) is called the *principle of angular impulse and momentum*. The quantities with the caret placed over the letter symbol denote the time integrals of the forces or the moments, respectively. For example (cf. (1.50)),

$$\hat{F}_x = \int_{t_0}^{t} F_x \, d\bar{t} \, .$$

When the body rotates about a fixed point, the principle of angular impulse and momentum (3.34b) may also be applied with respect to this point. In particular, we will use Equations (3.34a,b) to solve problems involving impacts (see Section 3.3.3).

Let us now determine the kinetic energy T of a rigid body in planar motion. If we choose the center of gravity C as the point of reference, the components of the velocity of an arbitrary particle of the body are given by $\dot{x} = \dot{x}_c - \omega\eta$ and $\dot{y} = \dot{y}_c + \omega\xi$ (see (3.29a)). Thus,

$$T = \frac{1}{2} \int v^2 \, dm = \frac{1}{2} \int (\dot{x}^2 + \dot{y}^2) \, dm$$

$$= \frac{1}{2} \left\{ (\dot{x}_c^2 + \dot{y}_c^2) \int dm - 2\dot{x}_c \omega \int \eta \, dm \right.$$

$$\left. + 2\dot{y}_c \omega \int \xi \, dm + \omega^2 \int (\xi^2 + \eta^2) \, dm \right\}.$$

The first moments $\int \xi \, dm$ and $\int \eta \, dm$ with respect to the center of mass are zero. Therefore, using $\dot{x}_c^2 + \dot{y}_c^2 = v_c^2$ and $\int (\xi^2 + \eta^2) \, dm = \Theta_C$, we obtain the kinetic energy:

$$T = \frac{1}{2} m v_c^2 + \frac{1}{2} \Theta_C \omega^2 . \tag{3.35}$$

It is composed of two parts: the *translational kinetic energy* $m v_c^2 / 2$ and the *rotational kinetic energy* $\Theta_C \omega^2 / 2$.

The work-energy theorem which was derived for particles and systems of particles is also valid for rigid bodies:

$$T - T_0 = U . \tag{3.36}$$

Here, U is the work done by the external forces (moments) during the motion of the body from an initial position ⓪ to an arbitrary position. If the external forces (moments) can be derived from a potential V, then with $U = -(V - V_0)$ we obtain from (3.36) the conservation of energy relation

$$T + V = T_0 + V_0 = \text{const} . \tag{3.37}$$

E3.13   **Example 3.13** A homogeneous slender rod hits the support A with velocity v as shown in Fig. 3.28a. At the instant of impact, the rod latches onto the support and starts to rotate about point A without rebounding.

Calculate the angular velocity of the rod immediately after the impact. Determine the loss of kinetic energy.

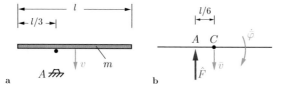

a   b   Fig. 3.28

Solution The changes of the velocities due to the impact are described by the principles of impulse and momentum (3.34a,b). The center of gravity of the rod has velocity v before the impact and the angular velocity of the rod is zero. Immediately after the impact we have the corresponding velocities $\bar{v}$ and $\dot{\bar{\varphi}}$, respectively. There are no horizontal components of the velocity or the impact force. Therefore, using the notation in Fig. 3.28b, we get

$$\downarrow:\quad m\bar{v} - mv = -\hat{F},$$

$$\overset{\curvearrowright}{C}:\quad \Theta_C\dot{\bar{\varphi}} = \frac{l}{6}\,\hat{F}.$$

The rod is still in the horizontal position immediately after the impact. Since it undergoes a rotation about point A, the kinematic relation between $\dot{\bar{\varphi}}$ and $\bar{v}$ is given by

$$\bar{v} = \frac{l}{6}\,\dot{\bar{\varphi}}.$$

We can solve these equations to obtain the angular velocity of the rod ($\Theta_C = ml^2/12$):

$$\underline{\underline{\dot{\bar{\varphi}} = \frac{3}{2}\frac{v}{l}}}.$$

It is also possible to calculate the angular velocity from only one equation. To do this we apply the principle of angular momentum and choose the fixed point A as the reference point. There exists no external moment with respect to A (the weight W can be neglected during the impact). Therefore the angular momentum is conserved:

$$\overset{\curvearrowright}{A}:\quad \frac{l}{6}mv = \Theta_A\dot{\bar{\varphi}} \;\rightarrow\; \dot{\bar{\varphi}} = \frac{lmv}{6\,\Theta_A} = \frac{lmv}{6\left[\dfrac{ml^2}{12} + m\left(\dfrac{l}{6}\right)^2\right]} = \frac{3}{2}\frac{v}{l}.$$

The loss of kinetic energy is the difference between the kinetic energies before the impact (pure translation) and after the impact (pure rotation about A):

$$\underline{\underline{\Delta T}} = T_0 - T = \frac{1}{2}mv^2 - \frac{1}{2}\Theta_A\dot{\bar{\varphi}}^2$$

$$= \frac{1}{2}mv^2 - \frac{1}{2}\left[\frac{ml^2}{12} + m\left(\frac{l}{6}\right)^2\right]\frac{9}{4}\frac{v^2}{l^2} = \frac{3}{8}mv^2 = \underline{\underline{\frac{3}{4}T_0}}\,.$$

E 3.14

Example 3.14 The center of a homogeneous wheel (mass m, radius r) has the velocity v_0 in the initial position as shown in Fig. 3.29. The wheel rolls without slipping.

Calculate the velocity v of its center in the position shown by the broken lines.

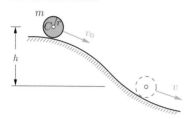

Fig. 3.29

Solution The only external force that acts on the system is the weight W of the wheel; it is a conservative force. Therefore conservation of energy immediately leads to the solution. With $v_0 = r\omega_0$ (the wheel rolls) and $\Theta_C = \frac{1}{2}mr^2$ we have

$$V_0 = 0, \quad T_0 = \frac{1}{2}mv_0^2 + \frac{1}{2}\Theta_C\omega_0^2 = \frac{3}{4}mv_0^2$$

in the initial position. In the final position (height difference h) we have

$$V = -mgh, \quad T = \frac{3}{4}mv^2\,.$$

Substituting into (3.37) yields

$$v = \underline{\underline{\sqrt{\frac{4}{3}gh + v_0^2}}}\,.$$

E 3.15

Example 3.15 A rope is wrapped around the circumference of a wheel (mass m_2, moment of inertia Θ_C), guided over an ideal pulley and attached to a block (mass m_1) as shown in Fig. 3.30a. The system is released from rest (unstretched spring).

Determine the velocity of the block as a function of the distance travelled. Assume that the wheel rolls without slipping and neglect the masses of the rope and pulley.

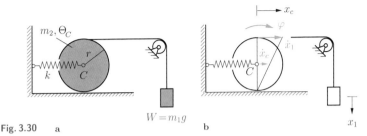

Fig. 3.30   a   b

Solution   We introduce the coordinates x_1, x_c and φ, measured from the initial position (Fig. 3.30b). In this position, the potential and the kinetic energy are zero:

$$V_0 = 0, \quad T_0 = 0.$$

Using the kinematic relations

$$x_c = r\varphi, \quad x_1 = 2\,x_c \quad \rightarrow \quad \dot{x}_c = r\dot{\varphi}, \quad \dot{x}_1 = 2\,\dot{x}_c$$

we obtain the energies in a displaced position:

$$V = -m_1\,g\,x_1 + \frac{k}{2}\,x_c^2 = -m_1\,g\,x_1 + \frac{k}{8}\,x_1^2,$$

$$T = \frac{1}{2}\,m_1\,\dot{x}_1^2 + \left(\frac{1}{2}\,m_2\,\dot{x}_c^2 + \frac{1}{2}\,\Theta_C\,\dot{\varphi}^2\right)$$

$$= \frac{1}{2}\,\dot{x}_1^2\left(m_1 + \frac{m_2}{4} + \frac{\Theta_C}{4\,r^2}\right).$$

Conservation of energy $T + V = T_0 + V_0$ (see (3.37)) leads to the unknown velocity of the block:

$$\dot{x}_1 = \pm\sqrt{\frac{2\,m_1\,g\,x_1 - \dfrac{k}{4}\,x_1^2}{m_1 + \dfrac{m_2}{4} + \dfrac{\Theta_C}{4\,r^2}}}\,.$$

In the special case of a vanishing numerator ($x_1 = 0$ or $x_1 = 8\,m_1\,g/k$), the velocity of the block is zero and the direction of the velocity reverses.

3.3.3 Eccentric Impact

In Section 2.5 we had to restrict ourselves to the investigation of problems involving central impact. The Equations (3.34a,b) allow us now to consider also problems where eccentric impact occurs (Fig. 3.31a). In such problems, the line connecting the centers of mass of the two bodies *does not* coincide with the normal to the plane of contact. In the following, we will always choose this normal (the line of impact) to be the x-axis.

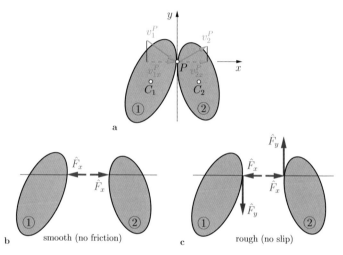

Fig. 3.31

A collision may be direct or oblique. We refer to a collision as being *direct*, if the velocities v_1^P and v_2^P of the points of contact P of both bodies have the direction of the line of impact immediately before the impact. An impact that is not direct is called *oblique*. If the surfaces of the bodies are *smooth*, the direction of the contact force coincides with the direction of the line of impact (Fig. 3.31b). If the surfaces are sufficiently *rough*, the bodies do not slip at point P during the impact. Then the contact force also has a component perpendicular to the line of impact in the case of an oblique collision (Fig. 3.31c).

The approach to a problem involving eccentric collision is analogous to the treatment in the case of a central collision. We have to apply the impulse laws (3.34a,b) to both bodies. In addition, we use the hypothesis (2.40) which we formally extend to an eccentric impact:

$$e = -\frac{\bar{v}_{1x}^P - \bar{v}_{2x}^P}{v_{1x}^P - v_{2x}^P} \, . \tag{3.38}$$

The quantities with a bar are the components of the velocities just after the impact (see Section 2.5). According to (3.38), the coefficient of restitution e is equal to the ratio of the relative velocity of separation (immediately after impact) to the relative velocity of approach (just before impact) of the contact points P on both bodies.

In the case of bodies with rough surfaces where no slipping occurs during the contact, we need an additional equation: the components of the velocities at P in the plane of impact (i.e., perpendicular to the line of impact) are equal during the collision and therefore also equal immediately after the collision:

$$\bar{v}_{1y}^P = \bar{v}_{2y}^P \, . \tag{3.39}$$

As an illustrative example of an eccentric and oblique impact we consider the collision of two bodies ① and ② with *smooth* surfaces (Fig. 3.32a). The masses m_1 and m_2 and the moments of inertia Θ_{C_1} and Θ_{C_2} are given. The velocities of the centers of mass and the angular velocities just before the collision are v_{1x}, v_{1y}, ω_1 and v_{2x}, v_{2y}, ω_2 (positive sense of rotation counterclockwise). Since the surfaces are smooth, the impulsive force has the direction of the line of impact. Using the notations given in Fig. 3.32b, the principles of impulse and momentum are for body ①

$$\rightarrow : \; m_1 \left(\bar{v}_{1x} - v_{1x} \right) = - \hat{F}_x, \quad \uparrow : \; m_1 \left(\bar{v}_{1y} - v_{1y} \right) = 0,$$
$$\curvearrowleft_{C_1} : \; \Theta_{C_1} \left(\bar{\omega}_1 - \omega_1 \right) = a_1 \, \hat{F}_x \tag{3.40a}$$

and for body ②

$$\rightarrow : \; m_2 \left(\bar{v}_{2x} - v_{2x} \right) = \hat{F}_x, \quad \uparrow : \; m_2 \left(\bar{v}_{2y} - v_{2y} \right) = 0,$$
$$\curvearrowleft_{C_2} : \; \Theta_{C_2} \left(\bar{\omega}_2 - \omega_2 \right) = - a_2 \, \hat{F}_x \, . \tag{3.40b}$$

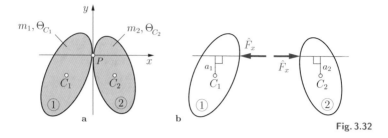

Fig. 3.32

In order to be able to apply (3.38)

$$e = -\frac{\bar{v}_{1x}^{P} - \bar{v}_{2x}^{P}}{v_{1x}^{P} - v_{2x}^{P}}$$

we have to write down the components of the velocities at point P in the direction of the line of impact just before and after the collision (cf. (3.29a)):

$$v_{1x}^{P} = v_{1x} - a_1\,\omega_1, \quad v_{2x}^{P} = v_{2x} - a_2\,\omega_2,$$

$$\bar{v}_{1x}^{P} = \bar{v}_{1x} - a_1\,\bar{\omega}_1, \quad \bar{v}_{2x}^{P} = \bar{v}_{2x} - a_2\,\bar{\omega}_2\,.$$

If the coefficient of restitution e is known, we now have 9 equations for the 9 unknowns (the velocities with a bar and $\hat{F}_x$). We can solve these equations to obtain, for example, the impulsive force $\hat{F}_x$:

$$\hat{F}_x = (1+e)\,\frac{v_{1x} - a_1\,\omega_1 - (v_{2x} - a_2\,\omega_2)}{\dfrac{1}{m_1} + \dfrac{1}{m_2} + \dfrac{a_1^2}{\Theta_{C_1}} + \dfrac{a_2^2}{\Theta_{C_2}}}\,.$$

Then the Equations (3.40a,b) yield the velocities of the centers of mass and the angular velocities immediately after the impact:

$$\bar{v}_{1x} = v_{1x} - \frac{\hat{F}_x}{m_1}, \quad \bar{v}_{1y} = v_{1y}, \quad \bar{\omega}_1 = \omega_1 + \frac{a_1\,\hat{F}_x}{\Theta_{C_1}},$$

$$\bar{v}_{2x} = v_{2x} + \frac{\hat{F}_x}{m_2}, \quad \bar{v}_{2y} = v_{2y}, \quad \bar{\omega}_2 = \omega_2 - \frac{a_2\,\hat{F}_x}{\Theta_{C_2}}\,.$$

Let us now consider a body that is pin-supported at A and therefore constrained to rotate about this point. It is struck by a linear impulse $\hat{F}$ (Fig. 3.33a). This generates a linear impulse reaction at A. All the impulses acting on the body are shown in the free-body diagram, Fig. 3.33b (the weight of the body can be neglected!). Assume that the body is at rest before the impact. Then the principles of impulse and momentum are given by

$$\nearrow : \quad m\,\bar{v}_x = \hat{F} - \hat{A}_x, \quad \nwarrow : \quad m\,\bar{v}_y = -\hat{A}_y,$$

$$\overset{\curvearrowleft}{A} : \quad \Theta_A\bar{\omega} = b\,\hat{F}.$$

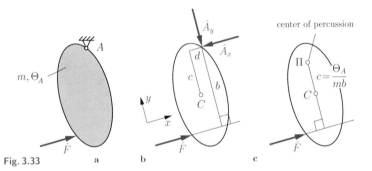

Fig. 3.33

With the velocities $\bar{v}_x = c\,\bar{\omega}$ and $\bar{v}_y = -d\,\bar{\omega}$ of the center of mass (rotation about the fixed point A) we obtain the impulsive pin reactions

$$\hat{A}_x = \hat{F}\left[1 - \frac{mcb}{\Theta_A}\right], \quad \hat{A}_y = \hat{F}\,\frac{mdb}{\Theta_A}.$$

These reactions are zero if we choose the position of the pin in such a way that it satisfies the conditions

$$c = \frac{\Theta_A}{mb} = \frac{r_{gA}^2}{b}, \quad d = 0. \tag{3.41}$$

Here, r_{gA} is the radius of gyration (see (3.18)). The point Π which is determined by (3.41) is referred to as the *center of percussion*. It lies on the straight line which is perpendicular to the direction of $\hat{F}$ and passes through C. Its distance from the center of mass is given by c (Fig. 3.33c). If the body is supported at this point, the impulsive reactions are zero. One makes use of this fact, for example, with a hammer or a tennis racket: the length of the handle is chosen in such a way that there are only small impulsive forces acting on the hand while striking a nail or a tennis ball, respectively. Note that the center of percussion coincides with the instantaneous center of rotation in the case that the body is not supported and therefore free to move in the plane immediately after the impact.

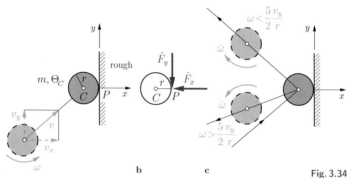

Fig. 3.34

We will now investigate a problem involving oblique impact and *rough* surfaces. A homogeneous sphere strikes a rough wall as shown in Fig. 3.34a. We assume that there is no slipping at the point of contact P during the collision. The velocities v_x, v_y of the center of mass and the angular velocity ω just before the impact are given; the corresponding velocities $\bar{v}_x$, $\bar{v}_y$, and $\bar{\omega}$ just after the impact are unknown. The principles of linear and angular impulse and momentum are

$$\rightarrow :\ m\left(\bar{v}_x - v_x\right) = -\hat{F}_x, \quad \uparrow:\ m\left(\bar{v}_y - v_y\right) = -\hat{F}_y,$$

$$\stackrel{\curvearrowleft}{C}:\ \Theta_C(\bar{\omega} - \omega) = -r\,\hat{F}_y$$

(cf. Fig. 3.34b). With $v_x^P = v_x$ and $\bar{v}_x^P = \bar{v}_x$, Equation (3.38) becomes

$$e = -\frac{\bar{v}_x^P}{v_x^P} = -\frac{\bar{v}_x}{v_x}.$$

There is no slipping at P during the collision. Therefore, using $\bar{v}_y^P = \bar{v}_y + r\bar{\omega}$ we find

$$\bar{v}_y^P = 0 \quad \rightarrow \quad \bar{v}_y + r\bar{\omega} = 0 \, .$$

Solving these equations and inserting $\Theta_C = \frac{2}{5}mr^2$ we obtain the velocities immediately after the impact:

$$\bar{v}_x = -e\,v_x, \quad \bar{v}_y = \dfrac{v_y - r\omega\,\dfrac{\Theta_C}{r^2\,m}}{1 + \dfrac{\Theta_C}{r^2\,m}} = \frac{5}{7}\,v_y - \frac{2}{7}\,r\omega,$$

$$\bar{\omega} = -\frac{\bar{v}_y}{r} = \frac{2}{7}\,\omega - \frac{5}{7}\,\frac{v_y}{r}\, .$$

Note that the velocity $\bar{v}_y$ is negative for $\omega > 5\,v_y/(2\,r)$ (Fig. 3.34c). In the case of $\omega < 5\,v_y/(2\,r)$, the sense of rotation is reversed during the impact ($\bar{\omega} < 0$).

Example 3.16 A homogeneous rod (mass $m_2 = 2m$) is initially at rest in the vertical position. It is struck by a particle (mass $m_1 = m$) with the velocity v (Fig. 3.35a). The coefficient of restitution e is given.

E3.16

Calculate the velocity of the particle and the angular velocity of the rod immediately after the impact.

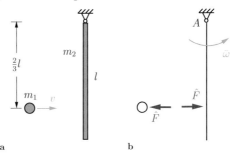

Fig. 3.35   a   b

Solution The particle performs a direct collision with the rod; the linear impulse acts along the line of impact (Fig. 3.35b). The principle of linear impulse and momentum for the particle is given by

$$\rightarrow:\quad m_1(\bar{v}-v)=-\hat{F}$$

and the principle of angular impulse and momentum for the rod $(\omega=0)$ reads

$$\overset{\curvearrowleft}{A}:\quad \Theta_A\,\bar{\omega}=\frac{2}{3}\,l\,\hat{F}\,.$$

The velocity of the rod at the point of impact just after the collision is $\frac{2}{3}l\bar{\omega}$. Thus, Equation (3.38) becomes

$$e=-\frac{\bar{v}-\frac{2}{3}l\bar{\omega}}{v}\,.$$

Now we have three equations for the three unknowns $\hat{F},\bar{v}$ and $\bar{\omega}$. Introduction of the given values of the masses and of $\Theta_A=m_2\,l^2/3$ yields the velocities

$$\underline{\underline{\bar{v}=\frac{v}{5}(2-3\,e)}},\quad \underline{\underline{\bar{\omega}=\frac{3}{5}\frac{v}{l}(1+e)}}\,.$$

In the case of an ideal plastic impact $(e=0)$ we obtain $\bar{v}=\frac{2}{5}v$ and $\bar{\omega}=\frac{3}{5}v/l$. An ideal elastic impact $(e=1)$ results in $\bar{v}=-v/5$ and $\bar{\omega}=\frac{6}{5}v/l$. There is no loss of kinetic energy in the case of an ideal elastic impact: $\bar{T}=\frac{1}{2}m\bar{v}^2+\frac{1}{2}\Theta_A\,\bar{\omega}^2=\frac{1}{2}mv^2=T$.

E3.17   **Example 3.17** Find the height h at which a homogeneous billiard ball (mass m) must be struck by a horizontal linear impulse (Fig. 3.36a) in order to have no sliding at the point of contact with the smooth horizontal surface.

Fig. 3.36

Solution There is no sliding at the point of contact A if the force $\hat{F}$ does not generate a horizontal linear impulse at A. Thus, this point has to be the center of percussion (= instantaneous center of rotation). With $c=r$ and $b=h$ (Fig. 3.36b), Equation (3.41) yields

$$r = \frac{\Theta_A}{mh}.$$

Since

$$\Theta_A = \Theta_C + mr^2 = \frac{2}{5} mr^2 + mr^2 = \frac{7}{5} mr^2$$

we obtain

$$\underline{\underline{h}} = \frac{\Theta_A}{mr} = \frac{7}{5} r.$$

Example 3.18   A homogeneous rod strikes a rough surface with velocity v as shown in Fig. 3.37a. Assume that the rod does not slide on the surface during the impact.

E3.18

Determine the velocity of the center of mass of the rod and the angular velocity just after the impact. Assume $e = 1$ (ideal elastic impact).

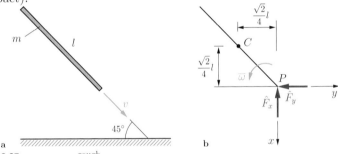

Fig. 3.37
a   rough   b

Solution   Since the rod does not slide at the point of contact, linear impulses in the x- and in the y-direction are exerted on the rod (Fig. 3.37b). With the velocities $v_x = v_y = v/\sqrt{2}$ of the center of mass and $\omega = 0$, the principles of impulse and momentum yield

$$\downarrow \ : \ m\left(\bar{v}_x - \frac{v}{\sqrt{2}}\right) = -\hat{F}_x, \tag{a}$$

$$\rightarrow : \ m\left(\bar{v}_y - \frac{v}{\sqrt{2}}\right) = -\hat{F}_y, \tag{b}$$

$$\overset{\curvearrowleft}{C}: \ \Theta_C \bar{\omega} = \frac{\sqrt{2}}{4} l \left(\hat{F}_x - \hat{F}_y\right). \tag{c}$$

The velocities of the point of contact P in the direction of the x-axis (line of impact) just before and after the collision are given by

$$v_x^P = v_x = \frac{v}{\sqrt{2}}, \quad \bar{v}_x^P = \bar{v}_x - \frac{\sqrt{2}}{4} l\bar{\omega}.$$

Then Equation (3.38) for an ideal elastic impact

$$e = -\frac{\bar{v}_x^P}{v_x^P} = -\frac{\bar{v}_x - \dfrac{\sqrt{2}}{4} l\bar{\omega}}{\dfrac{v}{\sqrt{2}}} = 1$$

leads to

$$\bar{v}_x - \frac{\sqrt{2}}{4} l\bar{\omega} = -\frac{v}{\sqrt{2}}. \tag{d}$$

The condition that there is no sliding at P during the collision yields

$$\bar{v}_y^P = 0 \quad \rightarrow \quad \bar{v}_y + \frac{\sqrt{2}}{4} l\bar{\omega} = 0. \tag{e}$$

With $\Theta_C = \dfrac{ml^2}{12}$, the unknown velocities follow from Equations (a) to (e):

$$\underline{\underline{\bar{v}_x = -\frac{5\sqrt{2}}{16} v}}, \quad \underline{\underline{\bar{v}_y = -\frac{3\sqrt{2}}{16} v}}, \quad \underline{\underline{\bar{\omega} = \frac{3v}{4l}}}.$$

3.4 Kinetics of a Rigid Body in Three Dimensional Motion

We will now discuss the equations which describe the three dimensional kinetics of a rigid body. The approach is analogous to the one we have used for the plane motion: by appropriate integration of Newton's second law for the motion of a particle we obtain the principles of linear and angular momentum. Here again we assume that the *internal* forces between the particles cancel out each other and therefore need not be considered (cf. Section 3.3.1).

3.4.1 Principles of Linear and Angular Momentum

Let us consider a rigid body with mass m which may be thought to consist of infinitesimal mass elements dm (Fig. 3.38). The external forces $d\boldsymbol{F}$ act on the elements while internal forces are assumed to cancel out. The location $\boldsymbol{r}_C$ of the center of mass C with respect to a *fixed* coordinate system x, y, z is given by (cf. Volume 1)

$$m\,\boldsymbol{r}_C = \int \boldsymbol{r}\,dm\,.$$

Differentiating this equation twice with respect to time, we obtain

$$m\,\dot{\boldsymbol{r}}_C = \int \dot{\boldsymbol{r}}\,dm, \tag{3.42}$$

$$m\,\ddot{\boldsymbol{r}}_C = \int \ddot{\boldsymbol{r}}\,dm\,. \tag{3.43}$$

The right-hand side of (3.42) represents the momentum $\boldsymbol{p}$ of the rigid body (sum of infinitesimal momenta). Thus, the momentum can be written as (cf. Section 2.2)

$$\boldsymbol{p} = m\,\boldsymbol{v}_C\,. \tag{3.44}$$

Since $\ddot{\boldsymbol{r}}\,dm = d\boldsymbol{F}$ and $\boldsymbol{F} = \int d\boldsymbol{F}$, the right-hand side of (3.43) represents the resultant of the external forces. Thus, the *principle of linear momentum* reads

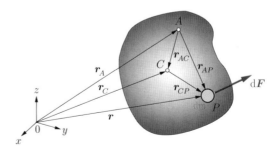

Fig. 3.38

$$m\,\ddot{\boldsymbol{r}}_C = \boldsymbol{F} \quad \text{or} \quad \dot{\boldsymbol{p}} = \boldsymbol{F}\,. \tag{3.45}$$

Accordingly, the motion of the center of mass is the same as the motion of a point mass m, acted upon by the resultant external force.

We will derive the principle of angular momentum only with respect to a *body-fixed* point A (Fig. 3.38). Taking the cross product of the equation of motion $\dot{\boldsymbol{v}}\,\mathrm{d}m = \mathrm{d}\boldsymbol{F}$ with $\boldsymbol{r}_{AP}$ and integration over the entire body yields

$$\int \boldsymbol{r}_{AP} \times \dot{\boldsymbol{v}}\,\mathrm{d}m = \int \boldsymbol{r}_{AP} \times \mathrm{d}\boldsymbol{F}\,. \tag{3.46}$$

The right-hand side represents the resultant moment $\boldsymbol{M}^{(A)}$ of the external forces with respect to point A. The left-hand side may be appropriately transformed by applying the identity

$$\boldsymbol{r}_{AP} \times \dot{\boldsymbol{v}} = (\boldsymbol{r}_{AP} \times \boldsymbol{v})^{\cdot} - \dot{\boldsymbol{r}}_{AP} \times \boldsymbol{v}\,.$$

Using the relations (cf. Section 3.1.3)

$$\boldsymbol{v} = \boldsymbol{v}_A + \boldsymbol{v}_{AP}, \quad \boldsymbol{v}_{AP} = \dot{\boldsymbol{r}}_{AP} = \boldsymbol{\omega} \times \boldsymbol{r}_{AP}$$

and

$$(\boldsymbol{\omega} \times \boldsymbol{r}_{AP}) \times (\boldsymbol{\omega} \times \boldsymbol{r}_{AP}) = \boldsymbol{0}, \quad \boldsymbol{r}_{AP} = \boldsymbol{r}_{AC} + \boldsymbol{r}_{CP}$$

it can be written as

$$\begin{aligned}
\boldsymbol{r}_{AP} \times \dot{\boldsymbol{v}} &= (\boldsymbol{r}_{AP} \times \boldsymbol{v})^{\cdot} - (\boldsymbol{\omega} \times \boldsymbol{r}_{AP}) \times [\boldsymbol{v}_A + (\boldsymbol{\omega} \times \boldsymbol{r}_{AP})] \\
&= (\boldsymbol{r}_{AP} \times \boldsymbol{v})^{\cdot} - (\boldsymbol{\omega} \times \boldsymbol{r}_{AP}) \times \boldsymbol{v}_A \\
&= (\boldsymbol{r}_{AP} \times \boldsymbol{v})^{\cdot} - [\boldsymbol{\omega} \times (\boldsymbol{r}_{AC} + \boldsymbol{r}_{CP})] \times \boldsymbol{v}_A \\
&= (\boldsymbol{r}_{AP} \times \boldsymbol{v})^{\cdot} - (\boldsymbol{\omega} \times \boldsymbol{r}_{AC}) \times \boldsymbol{v}_A - (\boldsymbol{\omega} \times \boldsymbol{r}_{CP}) \times \boldsymbol{v}_A\,.
\end{aligned}$$

Introducing now the *angular momentum* (sum of infinitesimal moments of momenta) with respect to A

$$\boldsymbol{L}^{(A)} = \int \boldsymbol{r}_{AP} \times \boldsymbol{v}\,\mathrm{d}m \tag{3.47}$$

and taking the relations $\int \boldsymbol{r}_{CP} \, \mathrm{d}m = \boldsymbol{0}$ (center of mass) and $\int \mathrm{d}m = m$ into account, the left-hand side of (3.46) leads to

$$
\int \boldsymbol{r}_{AP} \times \dot{\boldsymbol{v}} \, \mathrm{d}m = \frac{\mathrm{d}}{\mathrm{d}t} \int (\boldsymbol{r}_{AP} \times \boldsymbol{v}) \, \mathrm{d}m - (\boldsymbol{\omega} \times \boldsymbol{r}_{AC}) \times \boldsymbol{v}_A \int \mathrm{d}m
$$
$$
- (\boldsymbol{\omega} \times \int \boldsymbol{r}_{CP} \, \mathrm{d}m) \times \boldsymbol{v}_A
$$
$$
= \frac{\mathrm{d}\boldsymbol{L}^{(A)}}{\mathrm{d}t} - (\boldsymbol{\omega} \times \boldsymbol{r}_{AC}) \times \boldsymbol{v}_A \, m \, .
$$

Inserting this equation into (3.46) finally yields the *principle of angular momentum* in general form:

$$
\dot{\boldsymbol{L}}^{(A)} - (\boldsymbol{\omega} \times \boldsymbol{r}_{AC}) \times \boldsymbol{v}_A \, m = \boldsymbol{M}^{(A)} \, . \tag{3.48}
$$

Equation (3.48) takes a simpler form if either the center of mass C is chosen as the point of reference A ($\boldsymbol{r}_{AC} = \boldsymbol{0}$) or if the body-fixed point A coincides with a *fixed* in space point ($\boldsymbol{v}_A = \boldsymbol{0}$). In both cases the second term in (3.48) vanishes and the principle of angular momentum alternatively reads

$$
\dot{\boldsymbol{L}}^{(C)} = \boldsymbol{M}^{(C)} \text{ or } \dot{\boldsymbol{L}}^{(A)} = \boldsymbol{M}^{(A)}, \quad A \text{ fixed in space} \, . \tag{3.49}
$$

In words: the time rate of change of the moment of momentum (or angular momentum) is equal to the moment of the external forces.

3.4.2 Angular Momentum, Inertia Tensor, Euler's Equations

Introducing the velocity $\boldsymbol{v} = \boldsymbol{v}_A + \boldsymbol{\omega} \times \boldsymbol{r}_{AP}$ into (3.47), one obtains

$$
\boldsymbol{L}^{(A)} = \int \boldsymbol{r}_{AP} \, \mathrm{d}m \times \boldsymbol{v}_A + \int \boldsymbol{r}_{AP} \times (\boldsymbol{\omega} \times \boldsymbol{r}_{AP}) \, \mathrm{d}m \, .
$$

The first term on the right-hand side vanishes if we again choose the center of mass or a fixed in space point as our reference point A. The angular momentum then can be written as

$$
\boldsymbol{L}^{(A)} = \int \boldsymbol{r}_{AP} \times (\boldsymbol{\omega} \times \boldsymbol{r}_{AP}) \, \mathrm{d}m \, . \tag{3.50}
$$

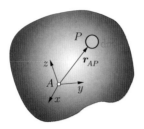

body-fixed coordinate system   **Fig. 3.39**

When $\boldsymbol{L}^{(A)}$ needs to be expressed explicitly, it is practical to use a body-fixed coordinate system x, y, z (Fig. 3.39). If we introduce

$$\boldsymbol{r}_{AP} = \begin{bmatrix} x \\ y \\ z \end{bmatrix}, \qquad \boldsymbol{\omega} = \begin{bmatrix} \omega_x \\ \omega_y \\ \omega_z \end{bmatrix} \tag{3.51}$$

and evaluate the cross products, (3.50) yields

$$\boldsymbol{L}^{(A)} = \begin{bmatrix} L_x^{(A)} \\ L_y^{(A)} \\ L_z^{(A)} \end{bmatrix} = \begin{bmatrix} \Theta_x\,\omega_x & + \Theta_{xy}\,\omega_y & + \Theta_{xz}\,\omega_z \\ \Theta_{yx}\,\omega_x & + \Theta_y\,\omega_y & + \Theta_{yz}\,\omega_z \\ \Theta_{zx}\,\omega_x & + \Theta_{zy}\,\omega_y & + \Theta_z\,\omega_z \end{bmatrix} \tag{3.52}$$

where

$$\Theta_x = \int (y^2 + z^2)\,\mathrm{d}m, \qquad \Theta_{xy} = \Theta_{yx} = -\int xy\,\mathrm{d}m,$$

$$\Theta_y = \int (z^2 + x^2)\,\mathrm{d}m, \qquad \Theta_{yz} = \Theta_{zy} = -\int yz\,\mathrm{d}m, \tag{3.53}$$

$$\Theta_z = \int (x^2 + y^2)\,\mathrm{d}m, \qquad \Theta_{zx} = \Theta_{xz} = -\int zx\,\mathrm{d}m.$$

The quantities $\Theta_x, \Theta_y, \Theta_z$ are the *moments of inertia* with respect to the x-, the y-, and the z-axes. They coincide with the moments of inertia which have been discussed in Section 3.2.2. The quantities Θ_{xy}, Θ_{yz} and Θ_{zx} are called the *products of inertia*.

Moments of inertia and products of inertia are components of the so-called *inertia tensor* $\boldsymbol{\Theta}^{(A)}$. They can be assembled in the following matrix:

$$\mathbf{\Theta}^{(A)} = \begin{bmatrix} \Theta_x & \Theta_{xy} & \Theta_{xz} \\ \Theta_{yx} & \Theta_y & \Theta_{yz} \\ \Theta_{zx} & \Theta_{zy} & \Theta_z \end{bmatrix} . \tag{3.54}$$

Since $\Theta_{xy} = \Theta_{yx}$ etc. the inertia tensor is symmetric. The inertial properties of the rigid body with respect to point A are uniquely described by $\mathbf{\Theta}^{(A)}$.

The moments of inertia and products of inertia (3.53) and consequently also the inertia tensor (3.54) depend on the chosen point of reference A as well as on the orientation of the axes x, y, and z. Without going into the details of the derivation it shall be noted that for each reference point there exists a specific coordinate system with three orthogonal axes 1, 2 and 3 for which all products of inertia are zero. These axes are referred to as *principal axes* (of inertia). The respective moments of inertia are extremal values and called *principal moments of inertia*. In the principal axes system the inertia tensor takes the simple form

$$\mathbf{\Theta}^{(A)} = \begin{bmatrix} \Theta_1 & 0 & 0 \\ 0 & \Theta_2 & 0 \\ 0 & 0 & \Theta_3 \end{bmatrix} \tag{3.55}$$

where Θ_1, Θ_2 and Θ_3 are the principal moments of inertia.

In the case of a homogeneous symmetric body, the symmetry axes are principal axes. Fig. 3.40a shows as an example the principal axes of a cuboid with respect to its center of mass. For axisymmetric bodies, the symmetry axis and any orthogonal axis are principal axes (Fig. 3.40b).

In the special case of a homogeneous thin plate, the mass of an element is given by $dm = \rho\, t\, dA$, see Fig. 3.40c. Since z is small compared with x and y, it can be neglected in the integration ($z \approx 0$). It then follows from (3.53) that the moment of inertia with respect to the x-axis can be written as

$$\Theta_x = \rho\, t \int y^2\, dA = \rho\, t\, I_x .$$

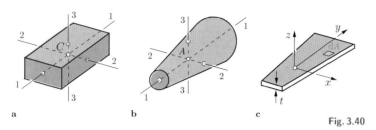

Fig. 3.40

Here, I_x is the area moment of inertia with respect to the x-axis (cf. Volume 2). Applying this procedure for all components of the inertia tensor, we obtain

$$\Theta_x = \rho\,t\,I_x, \qquad \Theta_y = \rho\,t\,I_y, \qquad \Theta_{xy} = \rho\,t\,I_{xy},$$
$$\Theta_z = \rho\,t\,(I_x + I_y) = \rho\,t\,I_p, \qquad \Theta_{xz} = \Theta_{yz} = 0. \tag{3.56}$$

Thus, in this case a direct relation exists between the mass moments of inertia and the area moments of inertia. Since the x, y-plane is a symmetry plane, one principal axis is orthogonal to the plane. The remaining two principal axes are located in the x, y-plane. Their orientation can be determined using the method valid for area moments of inertia.

We now return to the angular momentum as it is given by (3.52). With the inertia tensor (3.54) and the angular velocity vector according to (3.51) it can be represented as the matrix vector product:

$$\boldsymbol{L}^{(A)} = \boldsymbol{\Theta}^{(A)} \cdot \boldsymbol{\omega}. \tag{3.57}$$

This equation is the generalization of (3.14) to the three dimensional case. In mathematical terms it can be said: the angular momentum is a *linear vector function* of the angular velocity. Equation (3.57) can also be regarded as the mapping of the vector $\boldsymbol{\omega}$ onto the vector $\boldsymbol{L}^{(A)}$.

Generally, the angular momentum $\boldsymbol{L}^{(A)}$ and the angular velocity $\boldsymbol{\omega}$ of the body do *not* have the same direction. This can easily be seen from the representation (3.52). If, for example, we assume that $\boldsymbol{\omega}$ has the direction of the x-axis, then the components ω_y and ω_z are zero. In contrast, the y- and z-components of the angular momentum are only zero if the products of inertia vanish. As a consequence, $\boldsymbol{L}^{(A)}$ and $\boldsymbol{\omega}$ have the same direction only if the body rotates about a principal axis.

We will now introduce the angular momentum (3.57) into (3.49). When doing so we must take into account that the time derivative in (3.49) has to be taken with respect to a fixed in space (immovable) system while the angular momentum is given with respect to a body-fixed (moving) system. In Chapter 6 it will be shown that the relation between the time derivative $\mathrm{d}/\mathrm{d}t$ of a vector $\boldsymbol{L}$ with respect to a fixed in space system and the time derivative $\mathrm{d}^*/\mathrm{d}t$ with respect to a moving system is given by

$$\frac{\mathrm{d}\boldsymbol{L}}{\mathrm{d}t} = \frac{\mathrm{d}^*\boldsymbol{L}}{\mathrm{d}t} + \boldsymbol{\omega} \times \boldsymbol{L} \tag{3.58}$$

where $\boldsymbol{\omega}$ is the angular velocity of this system. Considering (3.58), we obtain from (3.49) and (3.57)

$$\boldsymbol{\Theta}^{(A)} \cdot \dot{\boldsymbol{\omega}} + \boldsymbol{\omega} \times (\boldsymbol{\Theta}^{(A)} \cdot \boldsymbol{\omega}) = \boldsymbol{M}^{(A)}. \tag{3.59}$$

Here A may be either the center of mass or a fixed in space point.

If the body-fixed system is chosen as the principal axes system where the inertia tensor takes the form (3.55), then (3.59) can be written in components as

$$\begin{aligned}
\Theta_1\, \dot{\omega}_1 \;-\; (\Theta_2 - \Theta_3)\, \omega_2\, \omega_3 &= M_1, \\
\Theta_2\, \dot{\omega}_2 \;-\; (\Theta_3 - \Theta_1)\, \omega_3\, \omega_1 &= M_2, \\
\Theta_3\, \dot{\omega}_3 \;-\; (\Theta_1 - \Theta_2)\, \omega_1\, \omega_2 &= M_3.
\end{aligned} \tag{3.60}$$

Here, M_1, M_2 and M_3 are the moments about the respective principal axes. Equations (3.60) are referred to as *Euler's equations*, named after the mathematician Leonhard Euler (1707–1783). This system of coupled nonlinear differential equations represents the principle of angular momentum with respect to a body-fixed principal axes system.

To solve Euler's equations may be difficult if the motion of the body-fixed system is *not* known a priori (e.g. gyroscope).

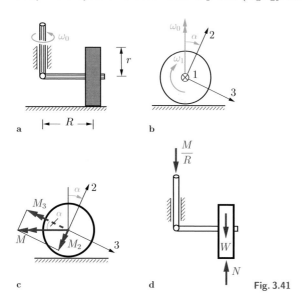

Fig. 3.41

As an illustrative example where this difficulty does not occur, we consider the rolling motion of a wheel in a grinding mill whose vertical king shaft rotates with constant angular velocity ω_0 (Fig. 3.41a). We assume that the moments of inertia with respect to the wheel-fixed rotating principal axes (see Fig. 3.41b) are known. Furthermore, since the wheel is assumed to be axisymmetric, $\Theta_2 = \Theta_3$. From kinematics we obtain

$$\omega_1 = \dot{\alpha} = \frac{R}{r}\,\omega_0, \qquad \dot{\omega}_1 = 0,$$

$$\omega_2 = \omega_0 \cos \alpha, \qquad \dot{\omega}_2 = -\,\omega_0\,\dot{\alpha}\sin\alpha = \omega_1\,\omega_3,$$

$$\omega_3 = -\,\omega_0 \sin \alpha, \qquad \dot{\omega}_3 = -\,\omega_0\,\dot{\alpha}\cos\alpha = -\,\omega_1\,\omega_2\,.$$

Thus, Euler's equations (3.60) lead to

$$M_1 = 0,$$

$$M_2 = (\Theta_2 - \Theta_3 + \Theta_1)\,\omega_1\,\omega_3 \;\; = \; -\,\Theta_1\,\omega_0^2\,\frac{R}{r}\,\sin\alpha,$$

$$M_3 = (-\Theta_3 - \Theta_1 + \Theta_2)\,\omega_1\,\omega_2 = -\,\Theta_1\,\omega_0^2\,\frac{R}{r}\,\cos\alpha\,.$$

Accordingly, the moment (couple) that acts on the wheel has the magnitude (see Fig. 3.41c)

$$M = \sqrt{M_2^2 + M_3^2} = \Theta_1\,\omega_0^2\,\frac{R}{r}\,.$$

Its direction is horizontal and perpendicular to the axis of the wheel. The compression force N between the wheel and the crushing bed is given by the sum of the weight W of the wheel and the gyroscopic part M/R (Fig. 3.41d):

$$N = W + \frac{M}{R} = W + \frac{\Theta_1\,\omega_0^2}{r}\,.$$

Therefore, the compression force can be considerably increased by increasing ω_0.

Example 3.19 Determine the moments and products of inertia with respect to the axes $\bar{x}, \bar{y}, \bar{z}$ and x, y, z for the homogeneous cuboid of mass m shown in Fig. 3.42.

E3.19

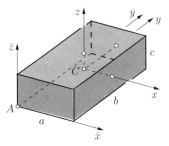

Fig. 3.42

Solution With $\mathrm{d}m = \rho\,\mathrm{d}\bar{x}\,\mathrm{d}\bar{y}\,\mathrm{d}\bar{z}$ and $m = \rho\,abc$ the quantities $\Theta_{\bar{x}}$ and $\Theta_{\bar{x}\bar{y}}$ are obtained as

$$\underline{\underline{\Theta_{\bar{x}}}} = \int (\bar{y}^2 + \bar{z}^2)\,\mathrm{d}m = \rho \int_0^c \int_0^b \int_0^a (\bar{y}^2 + \bar{z}^2)\,\mathrm{d}\bar{x}\,\mathrm{d}\bar{y}\,\mathrm{d}\bar{z} = \underline{\underline{\frac{m}{3}\,(b^2 + c^2)}},$$

$$\underline{\underline{\Theta_{\bar{x}\bar{y}}}} = -\int \bar{x}\bar{y}\,\mathrm{d}m = -\rho \int\limits_0^c \int\limits_0^b \int\limits_0^a \bar{x}\bar{y}\,\mathrm{d}\bar{x}\,\mathrm{d}\bar{y}\,\mathrm{d}\bar{z} = -\underline{\underline{\frac{mab}{4}}}.$$

Analogously we find

$$\underline{\underline{\Theta_{\bar{y}} = \frac{m}{3}(c^2 + a^2)}}, \quad \underline{\underline{\Theta_{\bar{z}} = \frac{m}{3}(a^2 + b^2)}}, \quad \underline{\underline{\Theta_{\bar{y}\bar{z}} = -\frac{mbc}{4}}}, \quad \underline{\underline{\Theta_{\bar{z}\bar{x}} = -\frac{mca}{4}}}.$$

Since x, y, z are axes of symmetry, they are principal axes. Therefore, the products of inertia $\Theta_{xy}, \Theta_{yz}, \Theta_{zx}$ are zero. The principal moments of inertia are given by

$$\underline{\underline{\Theta_1 = \Theta_x}} = 8\rho \int\limits_0^{c/2} \int\limits_0^{b/2} \int\limits_0^{a/2} (y^2 + z^2)\,\mathrm{d}x\,\mathrm{d}y\,\mathrm{d}z = \underline{\underline{\frac{m}{12}(b^2 + c^2)}},$$

$$\underline{\underline{\Theta_2 = \Theta_y = \frac{m}{12}(c^2 + a^2)}}, \quad \underline{\underline{\Theta_3 = \Theta_z = \frac{m}{12}(a^2 + b^2)}}.$$

For $c < a < b$ we find $\Theta_2 < \Theta_1 < \Theta_3$ and for $a = b = c$ (cube) all principal moments are equal: $\Theta_1 = \Theta_2 = \Theta_3 = ma^2/6$.

E3.20

Example 3.20 Determine the principal moments of inertia of a homogeneous circular shaft (radius r, length l, mass m) with respect to axes with their origin at the center of mass.

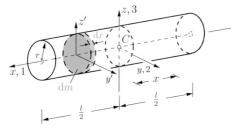

Fig. 3.43

Solution Since the body is axisymmetric, axes x, y, and z shown in Fig. 3.43 are the principal axes 1, 2, and 3. The moment of inertia about the x-axis has already been determined in Section 3.2.2:

$$\underline{\underline{\Theta_x = \Theta_1 = \frac{mr^2}{2}}}.$$

The moments of inertia about y and z are equal (symmetry).

To determine them we first calculate the moment of inertia of a circular disk having the mass dm and thickness dx (Fig. 3.43). Using the area moments of inertia $I_{y'} = I_{z'} = \pi r^4 / 4$ of a circular area (cf. Volume 2) we obtain

$$d\Theta_{y'} = d\Theta_{z'} = \rho\,dx\,I_{y'} = \rho\,\frac{\pi r^4}{4}\,dx\,.$$

Introducing $dm = \rho\pi r^2\,dx$, $m = \rho\pi r^2\,l$ and applying the parallel axis theorem (3.20) finally yields

$$\Theta_y = \Theta_z = \underline{\underline{\Theta_2 = \Theta_3}} = \int [d\Theta_{y'} + x^2\,dm]$$

$$= \int_{-l/2}^{+l/2} \left[\frac{1}{4}\,\rho\pi r^4\,dx + x^2\rho\pi r^2\,dx\right] = \underline{\underline{\frac{m}{12}(3\,r^2 + l^2)}}\,.$$

Example 3.21 A homogeneous circular cylinder (mass m, radius r, length l) rotates with constant angular velocity ω about a fixed axis having the angle α with respect to the cylinder axis (Fig. 3.44a). The center of mass C is located on the axis of rotation.

E3.21

Determine the support reactions.

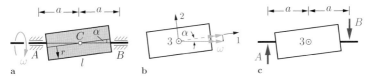

Fig. 3.44

Solution The center of mass C of the cylinder is at rest ($\ddot{\vec{r}}_C = \mathbf{0}$). According to (3.45) the resultant of the support reactions must vanish.

The moment of the support forces can be determined by using Euler's equations (3.60). For this purpose we introduce, according to Fig. 3.44b, the principal axes 1 and 2 (in the projection plane) and 3 (pointing outwards from the projection plane). The respective principal moments of inertia were determined in Example 3.20:

$$\Theta_1 = \frac{mr^2}{2}, \qquad \Theta_2 = \Theta_3 = \frac{m}{12}\left(l^2 + 3\,r^2\right).$$

With the components of the angular velocity vector (with respect to the principal axes)

$$\omega_1 = \omega \cos \alpha, \qquad \omega_2 = -\,\omega \sin \alpha, \qquad \omega_3 = 0$$

and $\dot{\omega}_1 = \dot{\omega}_2 = \dot{\omega}_3 = 0$ we obtain from (3.60)

$$M_1 = 0,$$
$$M_2 = 0,$$
$$M_3 = -(\Theta_1 - \Theta_2)\,\omega_1\,\omega_2 = \left[\frac{m}{2}\,r^2 - \frac{m}{12}\left(l^2 + 3\,r^2\right)\right]\omega^2 \sin \alpha \cos \alpha$$
$$= -\frac{m}{24}\left(l^2 - 3\,r^2\right)\omega^2 \sin 2\,\alpha\,.$$

Accordingly, the support forces must provide the couple M_3 about the third principal axis. Thus, with the notation according to Fig. 3.44c and $B = A$ (no resultant force) we obtain the support forces:

$$M_3 = -\,a\,A - a\,B = -\,2\,a\,A \quad \to \quad A = B = \frac{m\left(l^2 - 3\,r^3\right)\omega^2}{48\,a} \sin 2\,\alpha\,.$$

3.4.3 Support Reactions in Plane Motion

In Section 3.3 we discussed *plane motion* of a rigid body. We will now investigate, from a three dimensional point of view, the conditions for such a motion to take place. If the x, y-plane is chosen as the plane of motion, the angular velocity vector points in the z-direction: $\boldsymbol{\omega} = \omega\,\boldsymbol{e}_z$. Thus, we have

$$\omega_x = 0, \qquad \omega_y = 0, \qquad \omega_z = \omega.$$

In this case, from (3.59) with (3.45) the principle of angular momentum in component form reads

$$\Theta_{xz}\,\dot{\omega} - \Theta_{yz}\,\omega^2 = M_x\,,$$
$$\Theta_{yz}\,\dot{\omega} + \Theta_{xz}\,\omega^2 = M_y\,, \tag{3.61}$$
$$\Theta_z\,\dot{\omega} = M_z\,.$$

Here, the superscript A indicating the reference point (center of mass or fixed point) has been omitted.

The third equation of (3.61) represents the principle of angular momentum which was already derived in Section 3.3.1. The first and second equation of (3.61) indicate that moments M_x, M_y perpendicular to the z-axis must act if the products of inertia Θ_{xz}, Θ_{yz} are nonzero. Due to the principle actio = reactio, opposite moments are exerted on the supports.

An important application is the rotation of a body about a *fixed* axis. In this case, generally undesirable moments (couples) occur in the supports of technical systems (rotor, wheel). The rotating body is then referred to as *unbalanced*. The moments perpendicular to the rotation axis are only zero if the products of inertia vanish, i.e. if the rotation axis is a principal axis. During *dynamic balancing* the products of inertia of the body are brought to zero (or nearly zero) by affixing additional masses to the rotating body. A body is called *statically balanced* if its center of mass is located on the rotation axis.

Example 3.22 A thin homogeneous plate of mass m is free to rotate in supports A and B as shown in Fig. 3.45. Its rotation is driven by a constant moment M_0.

Formulate the equations of motion and determine the support reactions.

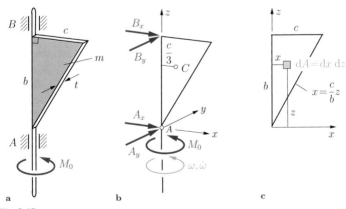

Fig. 3.45

Solution To describe the motion of the plate it would be sufficient to apply the principle of angular momentum about the fixed axis. However, to determine the support forces, the principles of angular momentum about the axes perpendicular to the rotation axis and the principle of linear momentum are needed.

We free the plate from the supports and choose a body-fixed (rotating) coordinate system x, y, z whose origin A is at rest (fixed in space). The center of mass rotates on a circle (Fig. 3.45b). With centripetal acceleration $a_{cx} = -c\omega^2/3$ and tangential acceleration $a_{cy} = c\dot{\omega}/3$ the components of the principle of linear momentum in the x- and in the y-direction read

$$m\,a_{cx} = A_x + B_x \quad \rightarrow \quad -\frac{mc\,\omega^2}{3} = A_x + B_x , \tag{a}$$

$$m\,a_{cy} = A_y + B_y \quad \rightarrow \quad \frac{mc\,\dot{\omega}}{3} = A_y + B_y . \tag{b}$$

To set up the components of the principle of angular momentum, the moments and products of inertia are needed. With $\mathrm{d}m = \rho\,t\,\mathrm{d}A$ and $m = \frac{1}{2}\rho\,t\,c\,b$ they are calculated as (cf. Fig. 3.45c)

$$\Theta_z = \rho\,t \int x^2\,\mathrm{d}A = \rho\,t \int_0^b \left\{ \int_0^{cz/b} x^2\,\mathrm{d}x \right\}\mathrm{d}z = \frac{mc^2}{6} ,$$

$$\Theta_{xz} = -\rho\,t \int xz\,\mathrm{d}A = -\rho\,t \int_0^b \left\{ \int_0^{cz/b} x\,\mathrm{d}x \right\}z\,\mathrm{d}z = -\frac{mcb}{4} ,$$

$$\Theta_{yz} = 0 .$$

Thus, from (3.61) we obtain

$$M_x = \Theta_{xz}\,\dot{\omega} \quad \rightarrow \quad -b\,B_y = -\frac{\dot{\omega}mcb}{4} , \tag{c}$$

$$M_y = \Theta_{xz}\,\omega^2 \quad \rightarrow \quad b\,B_x = -\frac{\omega^2 mcb}{4} , \tag{d}$$

$$M_z = \Theta_z\,\dot{\omega} \quad \rightarrow \quad M_0 = \frac{\dot{\omega}mc^2}{6} . \tag{e}$$

The last equation represents the equation of motion. Assuming the initial condition $\omega(0) = 0$ leads to

$$\dot{\omega} = \frac{6\,M_0}{mc^2} \quad \rightarrow \quad \omega = \frac{6\,M_0}{mc^2}\,t\,.$$

Insertion into (a) - (d) finally yields

$$A_x = -\frac{\omega^2 mc}{12}, \quad A_y = \frac{M_0}{2\,c}, \quad B_x = -\frac{\omega^2 mc}{4}, \quad B_y = \frac{3\,M_0}{2\,c}\,.$$

Example 3.23 A parasitic mass m_0 is attached to the wheel of a   E3.23
car with rotation axis z as shown in Fig. 3.46.

Determine the masses m_1 and m_2 that must be attached at
locations ① and ② in order to balance the wheel.

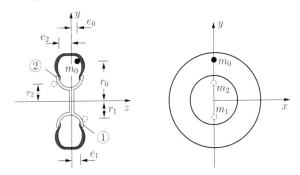

Fig. 3.46

Solution The wheel is balanced if the center of mass is located on
the rotation axis and if all the products of inertia vanish. Thus,
with the notation according to Fig. 3.46 the following conditions
must be fulfilled:

$$m_0\,r_0 + m_2\,r_2 = m_1\,r_1\,,$$

$$\Theta_{zy} = -\,m_0\,r_0\,e_0 + m_1\,r_1\,e_1 + m_2\,r_2\,e_2 = 0\,.$$

Solving yields the masses

$$m_1 = m_0\,\frac{r_0\,e_0 + e_2}{r_1\,e_1 + e_2}\,, \qquad m_2 = m_0\,\frac{r_0\,e_0 - e_1}{r_2\,e_1 + e_2}\,.$$

3.4.4 The Torque-Free Gyroscope

A rigid body rotating arbitrarily about its center of mass C or a
fixed in space point A (e.g. a pivot) is referred to as *gyroscope*.

We call a gyroscope *torque-free* if the moment of the external forces about the reference point C or A vanishes. This condition is fulfilled e.g. for the rotation of a satellite. Then, from (3.49) with $\boldsymbol{M}^{(C)} = \boldsymbol{0}$ or $\boldsymbol{M}^{(A)} = \boldsymbol{0}$ the conservation of angular momentum is obtained:

$$\boldsymbol{L}^{(C)} = \text{const} \qquad \text{or} \qquad \boldsymbol{L}^{(A)} = \text{const} . \tag{3.62}$$

In this case, Euler's equations (3.60) after inserting $M_i = 0$ read

$$\Theta_1 \dot{\omega}_1 - (\Theta_2 - \Theta_3) \omega_2 \omega_3 = 0,$$
$$\Theta_2 \dot{\omega}_2 - (\Theta_3 - \Theta_1) \omega_3 \omega_1 = 0, \tag{3.63}$$
$$\Theta_3 \dot{\omega}_3 - (\Theta_1 - \Theta_2) \omega_1 \omega_2 = 0 .$$

A special solution of (3.63) is the rotation with constant angular velocity ω_0 about one of the principal axes, e.g. the 1-axis: $\omega_1 = \omega_0 = \text{const}$, $\omega_2 = \omega_3 = 0$. In this case we have $\dot{\omega}_2 = \dot{\omega}_3 = 0$, i.e. the 1-axis is fixed in space.

Of practical interest is the question of how the system reacts to a *small perturbation* of the state of motion. To answer this question let us consider a neighboring state deviating only "slightly" from the initial state: $\omega_1 \approx \omega_0$, $\omega_2 \ll \omega_1$, $\omega_3 \ll \omega_1$. If we neglect the small product $\omega_2\omega_3$, then the first equation of (3.63) is approximately satisfied. The last two equations take the form

$$\Theta_2 \dot{\omega}_2 - (\Theta_3 - \Theta_1) \omega_0 \omega_3 = 0,$$
$$\Theta_3 \dot{\omega}_3 - (\Theta_1 - \Theta_2) \omega_0 \omega_2 = 0 .$$

Eliminating e.g. ω_3, we obtain an equation for ω_2:

$$\ddot{\omega}_2 + \lambda^2 \omega_2 = 0 \qquad \text{with} \qquad \lambda^2 = \omega_0^2 \frac{(\Theta_1 - \Theta_2)(\Theta_1 - \Theta_3)}{\Theta_2 \Theta_3} .$$

Its solution is given by (cf. also Section 5.2.1)

$$\omega_2(t) = \begin{cases} A_1 \cos \lambda t + B_1 \sin \lambda t & \text{for } \lambda^2 > 0, \\ A_2 \, e^{\lambda^* t} + B_2 \, e^{-\lambda^* t} & \text{for } \lambda^2 = -\lambda^{*2} < 0 . \end{cases}$$

For $\lambda^2 > 0$ the time dependent behavior of ω_2 (and also ω_3) is periodic; nevertheless, the perturbation remains bounded. Such a rotation is called *stable*. In contrast, for $\lambda^2 < 0$ the perturbation ω_2 increases exponentially with time. The motion then increasingly deviates from the initial state; this state is called *unstable*. Thus, a stable rotation about the principal axis 1 occurs for

$$(\Theta_1 - \Theta_2)(\Theta_1 - \Theta_3) > 0,$$

i.e. if the rotation axis coincides with the axis of the maximum principal moment ($\Theta_1 > \Theta_2, \Theta_3$) or with the axis of the minimum principal moment ($\Theta_1 < \Theta_2, \Theta_3$). A stable rotation about the axis of the intermediate principal moment ($\Theta_2 < \Theta_1 < \Theta_3$) is not possible.

As a special case we finally consider the *spherical gyrostat* ($\Theta_1 = \Theta_2 = \Theta_3$). Here, from (3.63) we directly obtain the result $\dot{\omega}_1 = \dot{\omega}_2 = \dot{\omega}_3 = 0$, i.e. the rotation is always stable.

3.5 Supplementary Examples

3.5

Detailed solutions to the following examples are given in **(A)** D. Gross et al. *Formeln und Aufgaben zur Technischen Mechanik 3*, Springer, Berlin 2010, or **(B)** W. Hauger et al. *Aufgaben zur Technischen Mechanik 1-3*, Springer, Berlin 2008.

Example 3.24 Point A of the rod in Fig. 3.47 moves with the constant velocity v_A to the left.

Determine the velocity and the acceleration of point B of the rod (point of contact with the step) as a function of the angle φ. Find the path $y(x)$ of the instantaneous center of rotation.

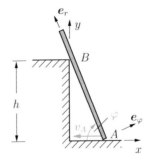

E3.24

Fig. 3.47

Results: see **(B)** $v_B = v_A \cos \varphi \, e_r$,

$$a_B = \frac{v_A^2}{h} \sin^2 \varphi (-\sin \varphi \, e_r + 2 \cos \varphi \, e_\varphi), \quad y = \frac{x^2}{h} + h.$$

E3.25

Example 3.25 The wheel of a crank drive rotates with constant angular velocity ω (Fig. 3.48).

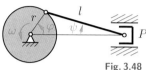

Determine the velocity and the acceleration of the piston P.

Fig. 3.48

Results: see (**A**) $v_P = -r\omega\left\{\sin\varphi + \dfrac{r}{l}\,\dfrac{\sin\varphi\cos\varphi}{\cos\psi}\right\}$,

$$a_P = -r\omega^2\left\{\cos\varphi - \dfrac{r}{l}\left[\dfrac{\sin^2\varphi}{\cos\psi} - \dfrac{\cos^2\varphi}{\cos^3\psi}\right]\right\}.$$

E3.26

Example 3.26 Link MA of the mechanism in Fig. 3.49 rotates with the angular velocity $\dot\varphi(t)$.

Determine the velocities of points B and C, the angular velocity ω and the angular acceleration $\dot\omega$ of the angled member ABC at the instant shown.

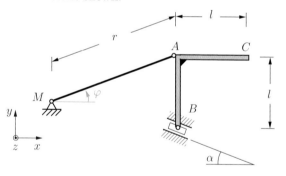

Fig. 3.49

Results: see (**B**) $\boldsymbol{v}_B = r\,\dot\varphi\,\dfrac{\cos\varphi}{\sin\alpha}\,[-\cos\alpha, \sin\alpha, 0]^T$,

$\boldsymbol{v}_C = r\,\dot\varphi\left[-\sin\varphi, \sin\varphi + \cos\varphi - \dfrac{\cos\varphi}{\tan\alpha}, 0\right]^T$,

$\omega = \dfrac{r\,\dot\varphi}{l}\left(\sin\varphi - \dfrac{\cos\varphi}{\tan\alpha}\right)$,

$\dot\omega = \dfrac{1}{l}\left[r\,\ddot\varphi\sin\varphi + r\dot\varphi^2\cos\varphi - \dfrac{1}{\tan\alpha}(r\,\ddot\varphi\cos\varphi - r\dot\varphi^2\sin\varphi + l\,\omega^2)\right]$.

Example 3.27 Wheel ① rolls in a gear mechanism without slip along circle ②. The mechanism is driven with a constant angular velocity Ω (Fig. 3.50).

Determine the magnitudes of the velocity and the acceleration of point P on the wheel.

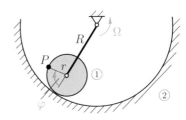

Fig. 3.50

E 3.27

Results: see (**A**)

$$v_P = 2\Omega R \sin \frac{\varphi}{2}, \qquad a_P = \Omega^2 R \sqrt{1 + \left(\frac{R}{r}\right)^2 + 2\frac{R}{r}\cos\varphi}.$$

Example 3.28 A disk ① (mass m, radius r_1) rests in a frictionless support ($\omega_0 = 0$). A second disk ② (mass m, radius r_2) rotates with the angular velocity ω_2. It is placed on disk ① as shown in Fig. 3.51. Due to friction, both disks eventually rotate with the same angular velocity $\bar{\omega}$.

Determine $\bar{\omega}$. Calculate the change ΔT of the kinetic energy.

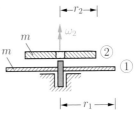

Fig. 3.51

E 3.28

Results: see (**B**) $\quad \bar{\omega} = \dfrac{r_2^2}{r_1^2 + r_2^2}\,\omega_2, \qquad \Delta T = -\dfrac{1}{4}\,m\,\omega_2^2\,\dfrac{r_1^2\,r_2^2}{r_1^2 + r_2^2}.$

Example 3.29 The door (mass m, moment of inertia Θ_A) of a car is open (Fig. 3.52). Its center of mass C has a distance b from the frictionless hinges. The car starts to move with the constant acceleration a_0.

Determine the angular velocity of the door when it bangs shut.

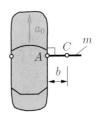

Fig. 3.52

E 3.29

Result: see (**B**) $\quad \dot{\varphi}(\frac{\pi}{2}) = \sqrt{\dfrac{2\,m\,a_0\,b}{\Theta_A}}.$

E3.30

Example 3.30 A child (mass m) runs along the rim of a circular platform (mass M, radius r) starting from point A (Fig. 3.53). The platform is initially at rest; its support is frictionless.

Determine the angle of rotation of the platform when the child arrives again at point A.

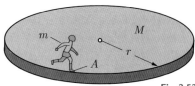

Fig. 3.53

Result: see (**B**) $\varphi = -\dfrac{2\pi}{M/(2m) + 1}$.

E3.31

Example 3.31 A homogeneous triangular plate of weight $W = mg$ is suspended from three strings with negligible mass.

Determine the acceleration of the plate and the forces in the strings just after string 3 is cut.

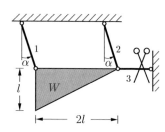

Fig. 3.54

Results: see (**A**) $a = g\sin\alpha$,

$$S_1 = \frac{mg}{6}\left(\sin\alpha + 4\cos\alpha\right), \quad S_2 = \frac{mg}{6}\left(-\sin\alpha + 2\cos\alpha\right).$$

E3.32

Example 3.32 A sphere (mass m_1, radius r) and a cylindrical wheel (mass m_2, radius r) are connected by two bars (mass of each bar $m_3/2$, length l). They roll down a rough inclined plane (with angle α) without slipping (Fig. 3.55).

Find the acceleration of the bars.

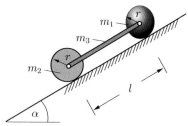

Fig. 3.55

Result: see (**B**) $a = \dfrac{(m_1 + m_2 + m_3)\sin\alpha}{7\,m_1/5 + 3\,m_2/2 + m_3}\,g$.

Example 3.33 The cylindrical shaft shown in Fig. 3.56 has a varying mass density given by $\rho = \rho_0(1 + \alpha r)$.

Find the moments of inertia Θ_x and Θ_y.

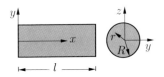

Fig. 3.56

E 3.33

Results: see (**A**) $\Theta_x = \dfrac{\pi}{2}\rho_0 l R^4\left(1 + \dfrac{4}{5}\alpha R\right),$

$$\Theta_y = \pi\rho_0 R^2 l\left[\frac{R^2}{4} + \frac{l^2}{3} + \alpha\left(\frac{R^3}{5} + \frac{2}{9}Rl^2\right)\right].$$

Example 3.34   Determine the moment of inertia Θ_a of a homogeneous torus with a circular cross section and mass m.

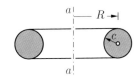

Fig. 3.57

E 3.34

Result: see (**A**) $\Theta_a = m\left(R^2 + \dfrac{3}{4}c^2\right).$

Example 3.35 A rope drum on a rough surface is set into motion by pulling the rope with a constant force F_0.

Determine the acceleration of point C assuming that the drum rolls (no slipping). What coefficient of static friction μ_0 is necessary to ensure rolling?

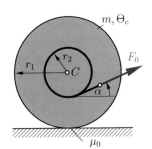

Fig. 3.58

E 3.35

Results: see (**A**)

$$a_c = \frac{F_0}{m}\frac{\cos\alpha - \dfrac{r_2}{r_1}}{1 + \dfrac{\Theta_c}{r_1^2 m}}, \qquad \mu_0 \geq \frac{\dfrac{\Theta_c\cos\alpha}{r_1^2 m} + \dfrac{r_2}{r_1}}{\left(\dfrac{mg}{F_0} - \sin\alpha\right)\left(1 + \dfrac{\Theta_c}{r_1^2 m}\right)}.$$

E3.36

Example 3.36 A homogeneous beam (mass M, length l) is initially in vertical position ① (Fig. 3.59). A small disturbance causes the beam to rotate about the frictionless support A (initial velocity equal to zero). In position ② it strikes a small sphere (mass m, radius $r \ll l$). Assume the impact to be elastic ($e = 1$).

Determine the angular velocities of the beam immediately before and after the impact and the velocity of the sphere after the impact.

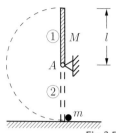

Fig. 3.59

Results: see (**B**) $\omega = \sqrt{6\,g/l}$, $\bar{\omega} = \dfrac{M - 3\,m}{M + 3\,m}\omega$, $\bar{v} = \dfrac{2\,l\,M}{M + 3\,m}\omega$.

E3.37

Example 3.37   A thin half-cylindrical shell of weight $W = mg$ rolls without sliding on a flat surface (Fig. 3.60).

Determine the angular velocity as a function of φ when the initial condition $\dot{\varphi}(0) = 0$ is given.

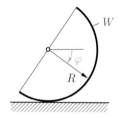

Fig. 3.60

Result: see (**A**) $\omega(\varphi) = \sqrt{\dfrac{2g \sin \varphi}{R(\pi - 2 \sin \varphi)}}$.

E3.38

Example 3.38   An elevator consists of a cabin (weight $W = mg$) which is connected through a rope (of negligible mass) with a rope drum and a band brake (coefficient of dynamic friction μ).

Determine the necessary braking force F such that a cabin travelling downwards with velocity v_0 stops after a distance h.

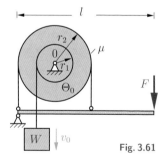

Fig. 3.61

Result: see (**A**) $F = \dfrac{r_1 v_0^2 m}{lh(1 - e^{-\mu\pi})}\left(1 + \dfrac{\Theta_0}{mr_1^2} + \dfrac{2gh}{v_0^2}\right)$.

Example 3.39 A homogeneous beam (mass m, length l) rotates about frictionless support A (Fig. 3.62) until it hits support C. The motion starts with zero initial velocity in the vertical position. The coefficient of restitution e is given.

Calculate the impulsive forces at A and C. Determine the distance a so that the impulsive force at A vanishes. Calculate the change of the kinetic energy.

E3.39

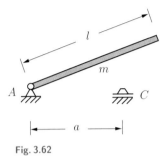

Fig. 3.62

Results: see (**B**) $\hat{A}_H = 0$, $\hat{A}_V = \dfrac{(1+e)(3a-2l)m}{6a}\sqrt{3gl}$,

$\hat{C} = \dfrac{(1+e)ml}{3a}\sqrt{3gl}$, $a = 2l/3$, $\Delta T = -(1-e^2)mgl/2$.

Example 3.40   A homogeneous circular disk of weight $W = mg$ is suspended from a pin-supported bar (of negligible mass). Initially the disk rotates with the angular velocity ω_0.
a) Determine the amplitude of oscillation of the pendulum, if the bar suddenly prevents the disk from rotating.
b) Calculate the energy loss ΔE.

E3.40

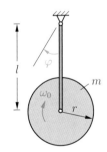

Fig. 3.63

Results: see (**A**)

$$\cos\varphi_1 = 1 - \frac{r^2\omega_0^2}{4gl}\frac{r^2}{r^2+2l^2}, \qquad \Delta E = \frac{mr^2\omega_0^2}{2}\frac{l^2}{r^2+2l^2}.$$

E3.41

Example 3.41   A homogeneous angled bar of mass m is attached to a
shaft with negligible mass. The rotation of the system is driven by the
moment M_0.

Determine the angular acceleration and the support reactions.

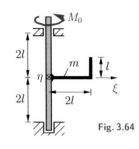

Fig. 3.64

Results: see **(A)** $\dot{\omega} = \dfrac{9M_0}{20ml^2}$,

$$A_\xi = -\frac{3}{4}\,ml\omega^2, \quad A_\eta = \frac{27}{80}\,\frac{M_0}{l}, \quad B_\xi = -\frac{7}{12}\,ml\omega^2, \quad B_\eta = \frac{21}{80}\,\frac{M_0}{l}.$$

E3.42

Example 3.42   A shaft (principal moments of inertia $\Theta_1, \Theta_2, \Theta_3$)
rotates with constant angular velocity ω_0 about its longitudinal
axis. This axis undergoes a rota-
tion $\alpha(t)$ about the z-axis of the
fixed in space system x, y, z.

Calculate the moment which
is exerted by the bearings on the
shaft for

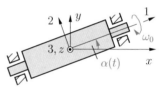

a) uniform rotation $\alpha = \Omega t$,

b) harmonic motion $\alpha = \alpha_0 \sin \Omega t$.

Fig. 3.65

Results: see **(A)**   a) $M_1 = M_3 = 0$, $M_2 = (\Theta_1 - \Theta_3)\omega_0\Omega$,

b) $M_1 = 0$, $M_2 = (\Theta_1 - \Theta_3)\omega_0\Omega\,\alpha_0 \cos \Omega t$, $M_3 = -\Theta_3\Omega^2\alpha_0 \sin \Omega t$.

E3.43

Example 3.43 A pin-supported rigid beam
(mass m, length l) is initially at rest. At
time $t_0 = 0$ it starts to rotate due to an
applied constant moment M_0.

Determine the stress resultants (internal forces and moments) as functions of x
for $t > t_0$. Neglect gravitational effects.

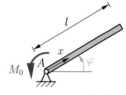

Fig. 3.66

Results: see **(A)** $M(x) = -M_0 \left(1 - \dfrac{x}{l}\right)^2 \left(1 + \dfrac{1}{2}\dfrac{x}{l}\right)$,

$$V(x) = \frac{3}{2}\frac{M_0}{l}\left[1 - \left(\frac{x}{l}\right)^2\right], \quad N(x) = \frac{9}{2}\frac{M_0^2\,t^2}{m\,l^3}\left[1 - \left(\frac{x}{l}\right)^2\right].$$

Example 3.44 The angled member (weight $W = mg$) in Fig. 3.67 consists of two homogeneous bars.

Derive the equation of motion of the vertical oscillations.

Result: see (**A**) $\ddot{\varphi} + \dfrac{\sqrt{65}}{14}\dfrac{g}{l}\sin\varphi = 0$

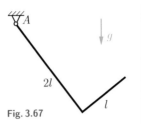

Fig. 3.67

E3.44

Example 3.45 A bowling ball (mass m) is placed on a rough surface (coefficient of kinetic friction $\mu = 0.3$) with the velocity $v_0 = 5$ m/s (Fig. 3.68). Initially, the ball does not rotate.

What is the position x_r of the ball when it stops to slip? Calculate the corresponding velocity v_r.

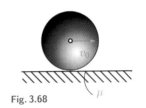

Fig. 3.68

Result: see (**A**) $x_r = 2.08$ m, $v_r = 3.57$ m/s.

E3.45

Example 3.46 The double pendulum in Fig. 3.69 consists of two equal homogeneous bars (each mass m, length l). It is struck by a linear impulse $\widehat{F}$ at point D.

Determine the distance d of point D from the lower end of the pendulum so that the angular velocity $\overline{\omega}_2$ of the lower bar is zero immediately after the impact. Calculate the impulsive forces at A and B.

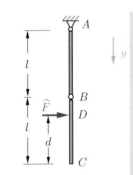

Fig. 3.69

Results: see (**A**) $d = \dfrac{5}{8}l$, $\widehat{A} = \dfrac{\widehat{F}}{8}$ and $\widehat{B} = \dfrac{\widehat{F}}{4}$.

E3.46

3.6 Summary

- The motion of a rigid body may be composed of translation and rotation; e.g. the velocity of a body-fixed point P is given by $\boldsymbol{v}_P = \boldsymbol{v}_A + \boldsymbol{\omega} \times \boldsymbol{r}_{AP}$.

  Plane motion at any instant may also be regarded as a pure rotation with angular velocity ω about the instantaneous center of rotation Π.

- During the rotation about a fixed axis, all body-fixed points undergo a circular motion with one and the same angular velocity $\omega = \dot{\varphi}$ and angular acceleration $\ddot{\varphi}$.

- Under the action of external forces a rigid body undergoes a motion which is described by the principle of linear momentum and the principle of angular momentum:

$$m\,\ddot{\boldsymbol{r}}_c = \boldsymbol{F}\,, \qquad \dot{\boldsymbol{L}}^{(A)} = \boldsymbol{M}^{(A)} \qquad (A = C \text{ or fixed}).$$

  In the plane case these reduce to three equations:

$$m\,\ddot{x}_c = F_x\,, \quad m\,\ddot{y}_c = F_y\,, \quad \Theta_A\,\ddot{\varphi} = M_A \quad \text{or} \quad \Theta_C\,\ddot{\varphi} = M_C\,.$$

- The kinetic energy of a rigid body is composed of translational and rotational energy. In the plane case it is given by

$$T = \tfrac{1}{2} m\,v_c^2 + \tfrac{1}{2}\Theta_C\,\omega^2\,.$$

- The impulse law and the work-energy theorem are analogous to those of a point mass or a system of point masses.

- When solving rigid body motion problems the following steps are usually necessary:
  ◇ Sketch of a free-body diagram containing all forces.
  ◇ Choice of coordinate system.
  ◇ Set-up of equations of motion. Reference point for principle of angular momentum is center of mass or a fixed point!
  ◇ Impact problems: set-up of impulse laws for the bodies involved and an impact hypothesis.
  ◇ Set-up of the necessary kinematical relations.
  ◇ Depending on the problem it might be advantageous to apply the work-energy theorem.

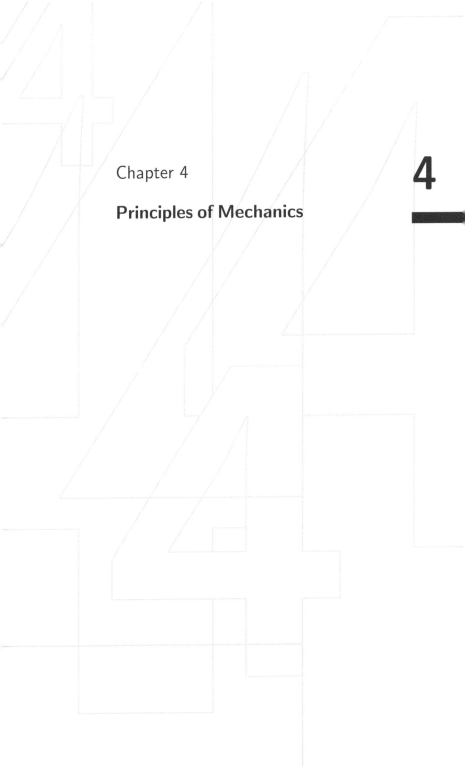

Chapter 4

Principles of Mechanics

4

4 Principles of Mechanics

Objectives: Until now we have used Newton's axioms to describe the motion of bodies. Equivalent are other basic laws which frequently are called *Principles of Mechanics*. In many cases it is of advantage to use these principles instead of Newton's laws when formulating the equations of motion. Several of them shall be discussed in this chapter and we will learn how to apply them.

4.1 Formal Reduction of Kinetics to Statics

According to Section 1.2.1, the motion of a *point mass* can be described by Newton's law of motion

$$m\boldsymbol{a} = \boldsymbol{F} \tag{4.1}$$

where $\boldsymbol{F}$ is the resultant of all forces acting on the point mass. We now rewrite (4.1) in the form

$$\boldsymbol{F} - m\boldsymbol{a} = \boldsymbol{0} \tag{4.2}$$

and consider the negative product of mass m and acceleration $\boldsymbol{a}$ formally as a force which after Jean Lerond d'Alembert (1717–1783) is referred to as *d'Alembert's inertial force* $\boldsymbol{F}_I$:

$$\boldsymbol{F}_I = -m\boldsymbol{a}\,. \tag{4.3}$$

This force is not a force in terms of Newton's force definition since *no* opposite force exists, i.e. the axiom of action and reaction is violated. Therefore we call it a *pseudo force* (*fictitious force*); its direction is opposite to the direction of the acceleration $\boldsymbol{a}$. Inserting (4.3) into (4.2), the law of motion can be written as

$$\boldsymbol{F} + \boldsymbol{F}_I = \boldsymbol{0}\,. \tag{4.4}$$

Accordingly, the motion of a point mass takes place such that the resultant $\boldsymbol{F}$ of the acting forces and d'Alembert's inertial force $\boldsymbol{F}_I$ are "in equilibrium". Since the point mass is not at rest but undergoes a motion, this is often referred to as a state of *dynamic equilibrium*.

By introducing the inertial force (4.3) we have *formally* reduced the law of motion (4.1) to the equilibrium condition (4.4). This procedure may be advantageous when setting up the equations of motion. If we want to apply this method to a problem, the inertial force $\boldsymbol{F}_I$ must be drawn into the free-body diagram in addition to the real forces. The equations of motion then are given by the condition "sum of all forces is equal to zero".

The plane motion of a *rigid body*, according to (3.31a,b), is described by the equations

$$m\ddot{x}_c = F_x, \qquad m\ddot{y}_c = F_y, \qquad \Theta_C\,\ddot{\varphi} = M_C. \qquad (4.5)$$

If in analogy to (4.3) the pseudo forces

$$F_{Ix} = -\,m\ddot{x}_c, \quad F_{Iy} = -\,m\ddot{y}_c \qquad (4.6)$$

and the pseudo moment

$$M_{IC} = -\,\Theta_C\,\ddot{\varphi} \qquad (4.7)$$

are introduced, the dynamic equilibrium conditions

$$F_x + F_{Ix} = 0, \quad F_y + F_{Iy} = 0, \quad M_C + M_{IC} = 0 \qquad (4.8)$$

are obtained. In the case of a pure rotation of a rigid body, the fixed point A may also be chosen as the reference point in the third equation of (4.8) instead of the center of mass C (cf. (3.33)).

If the motion of a *system* of rigid bodies is to be described with this method, the system must be cut into its individual bodies. Then, for each of the bodies the dynamic equilibrium conditions (4.4) or (4.8) can be written down.

E4.1

Example 4.1 A ship (mass m) has velocity v_0 after engine shutdown (Fig. 4.1a). The resistance force during gliding through the water is in good approximation given by $F_r = k\sqrt{v}$.

Assuming a straight path, determine the velocity as a function of time.

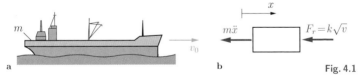

Fig. 4.1

Solution We introduce the coordinate x in the direction of the path, cf. Fig. 4.1b. The resistance force F_r acts on the vessel opposite to its direction of travel (in the vertical direction, the weight and the buoyancy are in equilibrium). According to the chosen x-

coordinate, a positive acceleration $\ddot{x}$ points to the right. The inertial force has the opposite direction (to the left) and the magnitude $m\ddot{x}$.

Dynamic equilibrium in the x-direction with $\ddot{x} = \dot{v}$ yields the equation of motion

$$\leftarrow : \quad m\dot{v} + k\sqrt{v} = 0 .$$

Separation of variables and integration lead in conjunction with the initial condition $v(0) = v_0$ to

$$\int_{v_0}^{v} \frac{\mathrm{d}\bar{v}}{\sqrt{\bar{v}}} = -\frac{k}{m} \int_0^t \mathrm{d}\bar{t} \quad \rightarrow \quad 2(\sqrt{v} - \sqrt{v_0}) = -\frac{k}{m} t .$$

Thus, the velocity as a function of time is given by

$$v = \left(\sqrt{v_0} - \frac{k}{2\,m} t \right)^2 .$$

Example 4.2 A point mass of weight $W = mg$ on a half-sphere (radius r) slips without friction downwards (Fig. 4.2a). The motion starts at the highest point with an initial velocity v_0.

At what location does the point mass lift-off from the sphere?

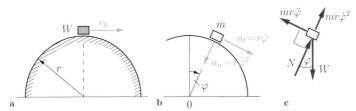

Fig. 4.2

Solution Until lift-off the point mass moves along a circular path; its location can be described by the coordinate φ (Fig. 4.2b). The point mass is subjected to its weight W (active force) and to the normal force N (reaction force). With the tangential acceleration $a_t = r\ddot{\varphi}$ (in the positive φ-direction) and the normal acceleration

$a_n = r\dot\varphi^2$ (directed towards 0) the inertial forces of magnitude ma_t (opposite to a_t) and ma_n (opposite to a_n) can be drawn in the free-body diagram (Fig. 4.2c).

Force equilibrium according to (4.4) in the normal and tangential directions yields the equations of motion

$$\nearrow:\quad N - mg\cos\varphi + mr\dot\varphi^2 = 0, \tag{a}$$

$$\searrow:\quad mg\sin\varphi - mr\ddot\varphi = 0. \tag{b}$$

We multiply (b) with $\dot\varphi$ and integrate to obtain

$$\dot\varphi\,\ddot\varphi = \frac{g}{r}\sin\varphi\,\dot\varphi \quad\rightarrow\quad \frac{1}{2}\dot\varphi^2 = -\frac{g}{r}\cos\varphi + C.$$

Using $v = r\dot\varphi$, the constant of integration C can be determined from the initial condition $v(\varphi = 0) = v_0$:

$$\frac{1}{2}\frac{v_0^2}{r^2} = -\frac{g}{r} + C \quad\rightarrow\quad C = \frac{1}{2}\frac{v_0^2}{r^2} + \frac{g}{r}.$$

Thus we have

$$\dot\varphi^2 = -\frac{2\,g}{r}\cos\varphi + \frac{v_0^2}{r^2} + \frac{2\,g}{r} = \frac{2\,g}{r}(1-\cos\varphi) + \frac{v_0^2}{r^2}.$$

Insertion into (a) yields the normal force N as a function of the angle φ:

$$N = mg\cos\varphi - mr\dot\varphi^2 = mg(3\cos\varphi - 2) - m\,\frac{v_0^2}{r}.$$

The location of lift-off is characterized by the condition of a vanishing normal force:

$$N = 0 \quad\rightarrow\quad mg\,(3\cos\varphi - 2) - m\,\frac{v_0^2}{r} = 0.$$

Thus, the lift-off angle φ^* is given by

$$\underline{\underline{\cos\varphi^* = \frac{2}{3} + \frac{v_0^2}{3\,gr}.}}$$

Since $\cos\varphi \leqq 1$ it follows that $v_0 \leqq \sqrt{gr}$. For $v_0 \geqq \sqrt{gr}$ the point mass immediately loses contact with the sphere.

Example 4.3 A block (mass m_1) is held by a rope which is guided over an ideal pulley (pulley and rope have negligible masses) and wrapped around a rope drum (mass m_2, moment of inertia Θ_C); see Fig. 4.3a.

E4.3

Find the equation of motion of the drum under the assumption that it rolls without slipping.

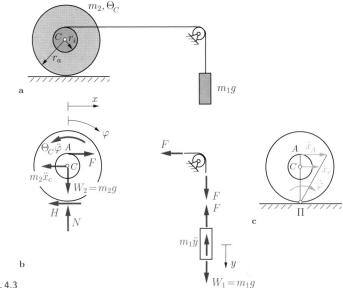

Fig. 4.3

Solution By appropriate section cuts we decompose the system into the individual bodies (Fig. 4.3b). To describe the motion, the coordinates x, φ (for the drum) and y (for the block) are introduced. The forces exerted upon the drum are its weight $W_2 = m_2\,g$, the normal force N, the static friction force H and the force F from the rope. The inertial force $m_2\ddot{x}_c$ points in negative x-direction and the pseudo moment $\Theta_C\,\ddot{\varphi}$ acts in the negative φ-direction (since the center of mass C moves in the x-direction, no inertial force in y-direction occurs). The block is subjected to its weight $W_1 = m_1\,g$ and the force F exerted by the rope. The inertial force $m_1\,\ddot{y}$ points in the negative y-direction.

Dynamic equilibrium yields for the drum

$$\rightarrow : \quad F - H - m_2\,\ddot{x}_c = 0\,,$$

$$\uparrow : \quad N - m_2\,g = 0\,,$$

$$\overset{\curvearrowright}{C} : \quad r_i F + r_a H - \Theta_C\,\ddot{\varphi} = 0$$

and for the block

$$\downarrow : \quad m_1\,g - F - m_1\,\ddot{y} = 0\,.$$

Since the drum is rolling (rotation about the instantaneous center of rotation Π) the following kinematic relation holds according to (3.10), cf. Fig. 4.3c:

$$\dot{x}_c = r_a\,\dot{\varphi} \quad \rightarrow \quad \ddot{x}_c = r_a\,\ddot{\varphi}\,.$$

The rope is considered as being inextensible. Therefore the velocities of the block and of point A are equal: $\dot{y} = \dot{x}_A$. Thus, with $\dot{x}_A = (r_i + r_a)\dot{\varphi}$ an additional kinematic relation is obtained:

$$\dot{y} = (r_i + r_a)\dot{\varphi} \quad \rightarrow \quad \ddot{y} = (r_i + r_a)\ddot{\varphi}\,.$$

Combining the equations yields the acceleration of the center of mass

$$\ddot{x}_c = \frac{m_1\,r_a\,(r_i + r_a)}{m_1\,(r_i + r_a)^2 + m_2\,r_a^2 + \Theta_C}\,g\,.$$

4.2 D'Alembert's Principle

When we investigate the motion of a body by applying Newton's law of motion or the dynamic equilibrium relation, we always obtain equations that contain all the forces acting on the body including the constraint forces. This procedure may be cumbersome for systems of point masses or bodies. Therefore, in this section we will become acquainted with a principle which leads to equations of motion that do *not* contain the constraint forces. This method is particularly advantageous when the constraint forces need not be determined. In the following we will restrict ourselves to motions where dry friction does not occur.

For simplicity, we will first consider the motion of a *point mass* on a prescribed path. According to (1.45) Newton's law reads

$$m\boldsymbol{a} = \boldsymbol{F}^{(a)} + \boldsymbol{F}^{(c)} \tag{4.9}$$

where $\boldsymbol{F}^{(a)}$ are the applied forces and $\boldsymbol{F}^{(c)}$ the constraint forces.

In order to obtain a formulation not containing the constraint or reaction forces, we use the notion of *virtual displacements*. These quantities are understood as fictitious, infinitesimal displacements that are kinematically admissible, i.e. consistent with the constraints of the system (cf. Volume 1, Section 8.2). Since the constraint forces $\boldsymbol{F}^{(c)}$ are perpendicular to the path and consequently perpendicular to the virtual displacements $\delta\boldsymbol{r}$, their virtual work vanishes:

$$\boldsymbol{F}^{(c)} \cdot \delta\boldsymbol{r} = 0\,. \tag{4.10}$$

This statement is referred to as *principle of d'Alembert*. It may be expressed in words as: the motion of a point mass takes place such that the virtual work of the constraint forces vanishes at all times.

Inserting (4.9) into (4.10) yields

$$(\boldsymbol{F}^{(a)} - m\boldsymbol{a}) \cdot \delta\boldsymbol{r} = 0\,. \tag{4.11}$$

Introducing the virtual work $\delta U = \boldsymbol{F}^{(a)} \cdot \delta\boldsymbol{r}$ of the applied forces $\boldsymbol{F}^{(a)}$ and the virtual work $\delta U_I = \boldsymbol{F}_I \cdot \delta\boldsymbol{r} = -m\boldsymbol{a} \cdot \delta\boldsymbol{r}$ for d'Alembert's inertial force $\boldsymbol{F}_I$, Equation (4.11) may be written as follows:

$$\delta U + \delta U_I = 0\,. \tag{4.12}$$

This form of the principle of d'Alembert is also referred to as *principle of virtual work*:

The motion of a point mass takes place such that the sum of the virtual works of the applied forces and d'Alembert's inertial force vanishes at all times.

Equation (4.12) does not contain the constraint forces any more.

The principle of virtual work (4.12) is also valid for a *system* of point masses with rigid constraints. To reveal this, Newton's equations of motion for a system of n point masses are written in analogy to (4.9) as (see also Chapter 2)

$$m_i \ddot{\boldsymbol{r}}_i = \boldsymbol{F}_i^{(a)} + \boldsymbol{F}_i^{(c)}, \quad i = 1, \dots, n. \tag{4.13}$$

During a virtual displacement of the system the total virtual work of the constraint forces vanishes:

$$\sum_i \boldsymbol{F}_i^{(c)} \cdot \delta \boldsymbol{r}_i = 0. \tag{4.14}$$

If the equations of motion (4.13) are dotted with $\delta \boldsymbol{r}_i$ and summed up over all point masses, we obtain with (4.14)

$$\sum_i (\boldsymbol{F}_i^{(a)} - m_i \ddot{\boldsymbol{r}}_i) \cdot \delta \boldsymbol{r}_i = 0. \tag{4.15}$$

Inserting $\sum_i \boldsymbol{F}_i^{(a)} \cdot \delta \boldsymbol{r}_i = \delta U$ and $\sum_i (-m_i \ddot{\boldsymbol{r}}_i) \cdot \delta \boldsymbol{r}_i = \delta U_I$ yields (4.12) again. The principle of virtual work (4.12) analogously is valid for *rigid bodies*.

If a system has several degrees of freedom, the number of independent virtual displacements is equal to the number of degrees of freedom. The principle of virtual displacements then yields just as many equations of motion as degrees of freedom exist.

E4.4

Example 4.4 Find the solution of Example 4.3 using the principle of virtual work.

Solution Since the constraint forces (normal force and static friction force at the drum, force in the rope) need not be calculated, the example may be treated with the principle of virtual

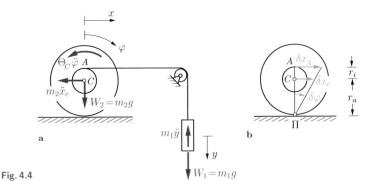

Fig. 4.4

work rather than with the method used in Example 4.3: freeing of the system then is *not* necessary!

The motion again is described by the coordinates x and φ for the drum and y for the block. The drum is subjected to the applied force $W_2 = m_2\, g$, the inertial force $m_2\, \ddot{x}_c$ and the pseudo moment $\Theta_C \ddot{\varphi}$. The weight $W_1 = m_1 g$ and the inertial force $m_1 \ddot{y}$ act at the block (Fig. 4.4a).

The system has *one* degree of freedom. For a virtual displacement the kinematic relations

$$\delta x_c = r_a\, \delta\varphi, \qquad \delta y = \delta x_A = (r_i + r_a)\, \delta\varphi \tag{a}$$

hold (Fig. 4.4b). Analogously we obtain (cf. Example 4.3)

$$\dot{x}_c = r_a\, \dot{\varphi}, \quad \dot{y} = \dot{x}_A = (r_i + r_a)\, \dot{\varphi}\,. \tag{b}$$

The virtual work of the applied forces and of the pseudo forces and moments are given by (cf. Fig. 4.4a)

$$\delta U = W_1\, \delta y = m_1\, g\, \delta y,$$
$$\delta U_I = -\, m_1\, \ddot{y}\, \delta y - m_2\, \ddot{x}_c \delta x_c - \Theta_C\, \ddot{\varphi}\, \delta\varphi\,. \tag{c}$$

Using (a) and (b), the variables x_c and y in (c) may be replaced by φ:

$$\delta U = m_1\, g\, (r_i + r_a)\, \delta\varphi,$$
$$\delta U_I = -\, m_1\, (r_i + r_a)^2 \ddot{\varphi}\, \delta\varphi - m_2\, r_a^2 \ddot{\varphi}\, \delta\varphi - \Theta_C \ddot{\varphi}\, \delta\varphi\,.$$

Thus, the principle of virtual work $\delta U + \delta U_I = 0$ yields

$$\{m_1\, g\, (r_i + r_a) - [\,m_1(r_i + r_a)^2 + m_2\, r_a^2 + \Theta_C]\, \ddot{\varphi}\}\, \delta\varphi = 0\,.$$

Since $\delta\varphi \neq 0$, it follows

$$[\,m_1\, (r_i + r_a)^2 + m_2\, r_a^2 + \Theta_C]\, \ddot{\varphi} = m_1\, g\, (r_i + r_a)$$

$$\rightarrow \quad \ddot{\varphi} = \frac{m_1\, (r_i + r_a)}{m_1\, (r_i + r_a)^2 + m_2\, r_a^2 + \Theta_C}\, g\,.$$

With $\ddot{x}_c = r_a\, \ddot{\varphi}$, the result of Example 4.3 is obtained again.

4.3 Lagrange Equations of the 2nd Kind

The equations of motion for a system of point masses may often be formulated in a substantially easier manner when specific coordinates are used. Then, by appropriately recasting the principle of virtual work (4.15), the so-called *Lagrange equations of the 2nd kind* are obtained. In the following we will derive these equations where we restrict ourselves to systems with either rigid constraints or internal forces which can be derived from a potential (e.g. the force of a spring).

According to (2.2), the number of degrees of freedom f of a system of n point masses in space, subjected to r kinematic constraints, is given by

$$f = 3\, n - r \qquad\qquad\qquad (4.16)$$

(in a plane we have $f = 2\, n - r$). Thus, the position of the system can be determined uniquely either by $3\, n$ (e.g. cartesian) coordinates which are linked by r constraint equations, or by f *independent* coordinates. These independent coordinates are called *generalized coordinates*.

As an example let us consider the simple pendulum as shown in Fig. 4.5. On the one hand we can specify the position of the mass m by the cartesian coordinates x and z. However, these coordinates are not independent of each other; they are linked by the constraint equation $x^2 + z^2 = l^2$. On the other hand, since the pendulum has *one* degree of freedom, its position can be specified by the *single* coordinate φ (= generalized coordinate). The relation between the cartesian coordinates and the generalized coordinate in our example is given by

$$x = l \sin \varphi, \quad z = l \cos \varphi\,.$$

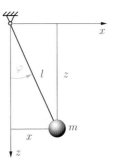

Fig. 4.5

The positions of the individual masses of a system of n point masses are described by the position vectors $\boldsymbol{r}_i$. The n position vectors $\boldsymbol{r}_i$ and the f generalized coordinates, denoted by q_j, then are related by

$$\boldsymbol{r}_i = \boldsymbol{r}_i(q_j), \quad i = 1, \ldots, n; \quad j = 1, \ldots, f\,. \tag{4.17}$$

To derive the Lagrange equations from (4.15), the virtual displacements $\delta\boldsymbol{r}_i$ are needed. According to (4.17), the n position vectors $\boldsymbol{r}_i$ depend, as just mentioned, on the f generalized coordinates q_j. Thus, the virtual displacements $\delta\boldsymbol{r}_i$ can be calculated analogously to the total differential of a function with several variables as follows:

$$\delta\boldsymbol{r}_i = \frac{\partial \boldsymbol{r}_i}{\partial q_1}\delta q_1 + \ldots + \frac{\partial \boldsymbol{r}_i}{\partial q_f}\delta q_f = \sum_j \frac{\partial \boldsymbol{r}_i}{\partial q_j}\delta q_j\,. \tag{4.18}$$

Inserting into (4.15) yields

$$\sum_i \left[(\boldsymbol{F}_i^{(a)} - m_i\ddot{\boldsymbol{r}}_i) \cdot \left(\sum_j \frac{\partial \boldsymbol{r}_i}{\partial q_j}\delta q_j \right) \right] = 0$$

and after multiplication we obtain

$$\sum_i \boldsymbol{F}_i^{(a)} \cdot \left(\sum_j \frac{\partial \boldsymbol{r}_i}{\partial q_j} \delta q_j \right) - \sum_i m_i \ddot{\boldsymbol{r}}_i \cdot \left(\sum_j \frac{\partial \boldsymbol{r}_i}{\partial q_j} \delta q_j \right) = 0. \quad (4.19)$$

Interchanging the order of the summation this can be written as

$$\sum_j \sum_i \boldsymbol{F}_i^{(a)} \cdot \frac{\partial \boldsymbol{r}_i}{\partial q_j} \delta q_j - \sum_j \sum_i m_i \ddot{\boldsymbol{r}}_i \cdot \frac{\partial \boldsymbol{r}_i}{\partial q_j} \delta q_j = 0 . \quad (4.20)$$

We now recast the second term in (4.20). For this purpose we use the identity

$$m_i \ddot{\boldsymbol{r}}_i \cdot \frac{\partial \boldsymbol{r}_i}{\partial q_j} = \frac{\mathrm{d}}{\mathrm{d}t} \left[m_i \dot{\boldsymbol{r}}_i \cdot \frac{\partial \boldsymbol{r}_i}{\partial q_j} \right] - m_i \dot{\boldsymbol{r}}_i \cdot \frac{\partial \dot{\boldsymbol{r}}_i}{\partial q_j} . \quad (4.21)$$

The correctness of this relation can be checked by differentiating the term in the square brackets.

Differentiation of (4.17) with respect to time leads to

$$\dot{\boldsymbol{r}}_i = \frac{\partial \boldsymbol{r}_i}{\partial q_1} \dot{q}_1 + \ldots + \frac{\partial \boldsymbol{r}_i}{\partial q_j} \dot{q}_j + \ldots + \frac{\partial \boldsymbol{r}_i}{\partial q_f} \dot{q}_f = \sum_j \frac{\partial \boldsymbol{r}_i}{\partial q_j} \dot{q}_j . \quad (4.22)$$

If we now differentiate with respect to $\dot{q}_j$, only the single term

$$\frac{\partial \dot{\boldsymbol{r}}_i}{\partial \dot{q}_j} = \frac{\partial \boldsymbol{r}_i}{\partial q_j} \quad (4.23)$$

of the sum remains. Thus, (4.21) takes the form

$$m_i \ddot{\boldsymbol{r}}_i \cdot \frac{\partial \boldsymbol{r}_i}{\partial q_j} = \frac{\mathrm{d}}{\mathrm{d}t} \left[m_i \dot{\boldsymbol{r}}_i \cdot \frac{\partial \dot{\boldsymbol{r}}_i}{\partial \dot{q}_j} \right] - m_i \dot{\boldsymbol{r}}_i \cdot \frac{\partial \dot{\boldsymbol{r}}_i}{\partial q_j}$$

$$= \frac{\mathrm{d}}{\mathrm{d}t} \left[\frac{\partial}{\partial \dot{q}_j} \left(\frac{1}{2} m_i \dot{\boldsymbol{r}}_i^2 \right) \right] - \frac{\partial}{\partial q_j} \left(\frac{1}{2} m_i \dot{\boldsymbol{r}}_i^2 \right) . \quad (4.24)$$

The kinetic energy of the system is given by

$$T = \sum_i \left(\frac{1}{2} m_i \dot{\boldsymbol{r}}_i^2 \right) . \quad (4.25)$$

Introducing the abbreviation

$$Q_j = \sum_i \boldsymbol{F}_i^{(a)} \cdot \frac{\partial \boldsymbol{r}_i}{\partial q_j} \quad (4.26)$$

and using (4.24)–(4.26) we obtain from (4.20)

$$\sum_j \left[Q_j - \frac{\mathrm{d}}{\mathrm{d}t} \left(\frac{\partial T}{\partial \dot{q}_j} \right) + \frac{\partial T}{\partial q_j} \right] \delta q_j = 0 . \quad (4.27)$$

As the generalized coordinates q_j are independent of each other so are the virtual displacements δq_j, i.e. they can be chosen arbitrarily. Therefore, the sum (4.27) vanishes only if each summand vanishes:

$$\frac{\mathrm{d}}{\mathrm{d}t}\left(\frac{\partial T}{\partial \dot{q}_j}\right) - \frac{\partial T}{\partial q_j} = Q_j, \quad j = 1, \ldots, f. \tag{4.28}$$

Equations (4.28) are called *Lagrange equations of the 2nd kind* after Joseph Louis Lagrange (1736-1813). They constitute a system of f equations for the f generalized coordinates q_j. In contrast, when using e.g. cartesian coordinates and Newton's laws we obtain $3\,n$ equations of motion and r constraint equations, i.e. in total $3\,n + r$ equations.

The total virtual work of the applied forces $\boldsymbol{F}_i^{(a)}$ is given by

$$\delta U = \sum_i \boldsymbol{F}_i^{(a)} \cdot \delta \boldsymbol{r}_i. \tag{4.29}$$

Replacing the virtual displacements $\delta \boldsymbol{r}_i$ according to (4.18) and using the abbreviation (4.26), it can be written as

$$\delta U = \sum_i \boldsymbol{F}_i^{(a)} \cdot \delta \boldsymbol{r}_i = \sum_i \boldsymbol{F}_i^{(a)} \cdot \left(\sum_j \frac{\partial \boldsymbol{r}_i}{\partial q_j} \delta q_j\right)$$

$$= \sum_j \sum_i \boldsymbol{F}_i^{(a)} \cdot \frac{\partial \boldsymbol{r}_i}{\partial q_j} \delta q_j = \sum_j Q_j \delta q_j. \tag{4.30}$$

Hence, the virtual work of the applied forces can be expressed by the quantities Q_j and the virtual displacements δq_j of the generalized coordinates. For this reason the quantities Q_j are referred to as *generalized forces*.

If the applied forces $\boldsymbol{F}_i^{(a)}$ can be derived from a potential V, the Lagrange equations (4.28) may be further simplified. Then we have (cf. (1.81))

$$\delta U = -\delta V. \tag{4.31}$$

The virtual change δV of the potential energy is calculated analogously to the total differential of a function of several variables:

$$\delta V(q_j) = \frac{\partial V}{\partial q_1} \delta q_1 + \ldots + \frac{\partial V}{\partial q_f} \delta q_f = \sum_j \frac{\partial V}{\partial q_j} \delta q_j . \tag{4.32}$$

By comparison of (4.30) and (4.32) we see

$$Q_j = -\frac{\partial V}{\partial q_j} \tag{4.33}$$

and insertion into (4.28) then yields

$$\frac{\mathrm{d}}{\mathrm{d}t} \left(\frac{\partial T}{\partial \dot{q}_j} \right) - \frac{\partial T}{\partial q_j} + \frac{\partial V}{\partial q_j} = 0 . \tag{4.34}$$

The potential energy V does not depend on $\dot{q}_j$. Therefore, if we introduce the *Lagrangian*

$$L = T - V \tag{4.35}$$

and take into account that $\partial V / \partial \dot{q}_j = 0$, we obtain from (4.34)

$$\frac{\mathrm{d}}{\mathrm{d}t} \left(\frac{\partial L}{\partial \dot{q}_j} \right) - \frac{\partial L}{\partial q_j} = 0, \quad j = 1, \ldots, f . \tag{4.36}$$

These are the Lagrange equations of the 2nd kind for conservative systems. Here they have been derived only for systems of point masses but they are analogously also valid for rigid bodies. They have the advantage that only the kinetic and potential energies must be set up. The equations of motion then follow simply by differentiation.

E4.5   **Example 4.5**   A point mass under the action of gravity moves without friction along a path having the shape of a parabola (Fig. 4.6a)

Derive the equation of motion.

Solution   The equation of the parabola is given in cartesian coordinates by $y = c\,x^2$ (Fig. 4.6b).

The applied force (weight) is conservative. The system has *one* degree of freedom; we choose the cartesian coordinate x as the generalized coordinate q.

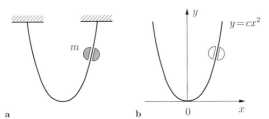

Fig. 4.6   a   b   0   x

The kinetic energy of the point mass is

$$T = \frac{1}{2} m v^2 = \frac{1}{2} m \left(\dot{x}^2 + \dot{y}^2 \right).$$

With

$$y = c x^2 \quad \rightarrow \quad \dot{y} = 2 c x \dot{x}$$

it takes the form

$$T = \frac{1}{2} m \left(\dot{x}^2 + 4 c^2 x^2 \dot{x}^2 \right).$$

If we choose the zero-level of the potential energy at the vertex of the parabola, we obtain

$$V = mg y = mgc x^2.$$

Thus, the Lagrangian (4.35) is given by

$$L = T - V = \frac{1}{2} m \left(\dot{x}^2 + 4 c^2 x^2 \dot{x}^2 \right) - mgc x^2.$$

Calculation of the derivatives

$$\frac{\partial L}{\partial \dot{x}} = m\dot{x} + 4 m c^2 x^2 \dot{x},$$

$$\frac{\mathrm{d}}{\mathrm{d}t} \left(\frac{\partial L}{\partial \dot{x}} \right) = m\ddot{x} + 8 m c^2 x \dot{x}^2 + 4 m c^2 x^2 \ddot{x},$$

$$\frac{\partial L}{\partial x} = 4 m c^2 x \dot{x}^2 - 2mgc x$$

and insertion into the Lagrange equation (4.36)

$$\frac{\mathrm{d}}{\mathrm{d}t} \left(\frac{\partial L}{\partial \dot{x}} \right) - \frac{\partial L}{\partial x} = 0$$

yields

$$\ddot{x}\left(1 + 4\,c^2\,x^2\right) + 4\,c^2\,x\,\dot{x}^2 + 2\,gc\,x = 0\,.$$

Example 4.6 A simple pendulum (length l, mass m_2) pivots on a block (mass m_1) as shown in Fig. 4.7a. The block is conntected to a wall by a spring (stiffness k) and can glide without friction on the support.

Find the equations of motion.

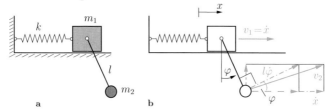

Fig. 4.7

Solution The system is conservative. Its position is uniquely determined by the displacement x of the block (measured from the equilibrium position of the unstretched spring) and the angle φ (Fig. 4.7b). Hence, the system has two degrees of freedom. We choose

$$q_1 = x, \quad q_2 = \varphi$$

as generalized coordinates. The kinetic energy is given by

$$T = \frac{1}{2}\,m_1\,v_1^2 + \frac{1}{2}\,m_2\,v_2^2\,.$$

The velocity of mass m_1 is $v_1 = \dot{x}$ whereas the velocity of mass m_2 is determined by the translation of mass m_1 and the superimposed rotation of the pendulum. According to Fig. 4.7b we obtain $v_2^2 = (\dot{x} + l\dot{\varphi}\cos\varphi)^2 + (l\dot{\varphi}\sin\varphi)^2$. Thus the kinetic energy can be written as

$$T = \frac{1}{2}\,m_1\,\dot{x}^2 + \frac{1}{2}\,m_2\left[(\dot{x} + l\dot{\varphi}\cos\varphi)^2 + (l\dot{\varphi}\sin\varphi)^2\right]. \tag{a}$$

If the zero-level of the potential of the weight is chosen at the level of the block, the total potential energy of the system is

$$V = \frac{1}{2} k \, x^2 - m_2 \, gl \cos \varphi \, . \tag{b}$$

With (a) and (b) the Lagrangian is given as

$$L = T - V = \frac{1}{2}(m_1 + m_2) \, \dot{x}^2 + m_2 \, l\dot{x}\dot{\varphi} \cos \varphi$$
$$+ \frac{1}{2} \, m_2 \, l^2 \, \dot{\varphi}^2 - \frac{1}{2} \, k \, x^2 + m_2 \, gl \cos \varphi .$$

To set up the Lagrange equations (4.36)

$$\frac{\mathrm{d}}{\mathrm{d}t} \left(\frac{\partial L}{\partial \dot{x}} \right) - \frac{\partial L}{\partial x} = 0, \qquad \frac{\mathrm{d}}{\mathrm{d}t} \left(\frac{\partial L}{\partial \dot{\varphi}} \right) - \frac{\partial L}{\partial \varphi} = 0 \tag{c}$$

the following derivatives must be calculated:

$$\frac{\partial L}{\partial \dot{x}} = (m_1 + m_2)\dot{x} + m_2 \, l\dot{\varphi} \cos \varphi,$$

$$\frac{\mathrm{d}}{\mathrm{d}t} \left(\frac{\partial L}{\partial \dot{x}} \right) = (m_1 + m_2)\ddot{x} + m_2 \, l\ddot{\varphi} \cos \varphi - m_2 \, l\dot{\varphi}^2 \sin \varphi,$$

$$\frac{\partial L}{\partial \dot{\varphi}} = m_2 \, l\dot{x} \cos \varphi + m_2 \, l^2 \dot{\varphi},$$

$$\frac{\mathrm{d}}{\mathrm{d}t} \left(\frac{\partial L}{\partial \dot{\varphi}} \right) = m_2 \, l\ddot{x} \cos \varphi - m_2 \, l\dot{x}\dot{\varphi} \sin \varphi + m_2 \, l^2\ddot{\varphi},$$

$$\frac{\partial L}{\partial x} = - \, k \, x, \qquad \frac{\partial L}{\partial \varphi} = - \, m_2 \, l\dot{x}\dot{\varphi} \sin \varphi - m_2 gl \, \sin \varphi \, .$$

Insertion into (c) yields the equations of motion

$$\underline{\underline{(m_1 + m_2) \, \ddot{x} + m_2 \, l\ddot{\varphi} \cos \varphi - m_2 \, l\dot{\varphi}^2 \sin \varphi + k \, x = 0,}}$$

$$\underline{\underline{\ddot{x} \cos \varphi + l\ddot{\varphi} + g \sin \varphi = 0 \, .}}$$

In the limit case $k \to \infty$, the first equation leads to $x = 0$ whereas the second one is reduced to the equation $l\ddot{\varphi} + g \sin \varphi = 0$ of the simple pendulum (cf. Section 5.2.1).

Example 4.7 The oscillator shown in Fig. 4.8a consists of a spring (stiffness k) and a point mass of weight $W = mg$. The length of the unstretched spring is l_0.

Determine the equations of motion. Assume that the oscillator moves in a plane.

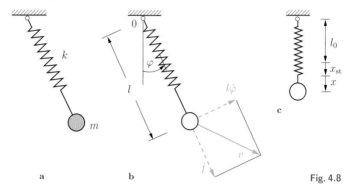

Fig. 4.8

Solution The position of the point mass is uniquely determined by its distance l from pin 0 and by the angle φ (Fig. 4.8b). Hence, the system has *two* degrees of freedom. It is often advantageous to use dimensionless quantities as generalized coordinates; here we choose

$$q_1 = l/l_0, \quad q_2 = \varphi. \tag{a}$$

The kinetic energy of the point mass is given by (cf. Fig. 4.8b)

$$T = \frac{1}{2} mv^2 = \frac{1}{2} m \left(\dot{l}^2 + l^2 \dot{\varphi}^2 \right). \tag{b}$$

Introducing (a) into (b), this is written as

$$T = \frac{1}{2} ml_0^2 (\dot{q}_1^2 + q_1^2 \dot{q}_2^2).$$

If we choose the zero-level of the potential energy for the weight W at pin 0, then the total potential energy of the system is given by

$$\begin{aligned}
V &= \frac{1}{2} k (l - l_0)^2 - mg\, l \cos \varphi \\
&= \frac{1}{2} k l_0^2 (q_1 - 1)^2 - mg\, l_0\, q_1 \cos q_2.
\end{aligned}$$

Thus, the Lagrangian of the conservative system can be expressed as

$$L = T - V = \frac{1}{2} m l_0^2 \, (\dot{q}_1^2 + q_1^2 \, \dot{q}_2^2) - \frac{1}{2} k \, l_0^2 \, (q_1 - 1)^2 + mg \, l_0 \, q_1 \cos q_2 \, .$$

To set up the Lagrange equations, the following derivatives are needed:

$$\frac{\partial L}{\partial \dot{q}_1} = m l_0^2 \, \dot{q}_1 , \qquad \frac{\mathrm{d}}{\mathrm{d}t} \left(\frac{\partial L}{\partial \dot{q}_1} \right) = m l_0^2 \, \ddot{q}_1 ,$$

$$\frac{\partial L}{\partial \dot{q}_2} = m l_0^2 \, q_1^2 \, \dot{q}_2 , \qquad \frac{\mathrm{d}}{\mathrm{d}t} \left(\frac{\partial L}{\partial \dot{q}_2} \right) = m l_0^2 \, (2 \, q_1 \, \dot{q}_1 \, \dot{q}_2 + q_1^2 \, \ddot{q}_2) ,$$

$$\frac{\partial L}{\partial q_1} = m \, l_0^2 \, q_1 \, \dot{q}_2^2 - k \, l_0^2 (q_1 - 1) + mg \, l_0 \cos q_2 ,$$

$$\frac{\partial L}{\partial q_2} = - mg \, l_0 \, q_1 \sin q_2 \, .$$

From (4.36) we then obtain the coupled equations of motion

$$\underline{\underline{m l_0 \, \ddot{q}_1 - m l_0 \, q_1 \, \dot{q}_2^2 + k l_0 (q_1 - 1) - mg \cos q_2 = 0}} \, , \tag{c}$$

$$\underline{\underline{l_0 \, q_1 \, \ddot{q}_2 + 2 \, l_0 \, \dot{q}_1 \, \dot{q}_2 + g \sin q_2 = 0}} \, .$$

In the special case when the rotation is prevented and the point mass moves along the vertical axis ($q_2 \equiv 0$), the second equation is fulfilled and the first one reduces to

$$m l_0 \, \ddot{q}_1 + k \, l_0 \, (q_1 - 1) - mg = 0$$

$$\rightarrow \quad m \ddot{l} + k \, l - k \, l_0 - mg = 0 \, . \tag{d}$$

If a new coordinate x with its origin at the equilibrium position is introduced (Fig. 4.8c), we have $l = l_0 + x_{st} + x$ where x_{st} is the static change of length of the spring. Then we obtain from (d) with $x_{st} = mg/k$:

$$m \ddot{x} + k \, l_0 + k \, x_{st} + k \, x - k \, l_0 - mg = 0 \quad \rightarrow \quad m \ddot{x} + k \, x = 0 \, .$$

This is the differential equation of a harmonic (spring-mass) oscillator (cf. Section 5.2.1).

In another special case, when $k \to \infty$, the first equation of motion in (c) leads to $q_1 = 1$, i.e. $l = l_0$. The second equation reduces to the equation of motion of a simple pendulum:

$$l_0 \ddot{q}_2 + g \sin q_2 = 0 \quad \to \quad \ddot{\varphi} + \frac{g}{l} \sin \varphi = 0.$$

4.4 Supplementary Examples

Detailed solutions to the following examples are given in (**A**) D. Gross et al. *Formeln und Aufgaben zur Technischen Mechanik 3*, Springer, Berlin 2010, or (**B**) W. Hauger et al. *Aufgaben zur Technischen Mechanik 1-3*, Springer, Berlin 2008.

E4.8

Example 4.8 A homogeneous disk (mass m, radius r) rolls without slipping on a rough surface (Fig. 4.9). Its center of mass C is connected with the wall by a spring (spring constant k).

Derive the equation of motion using

a) Newton's 2nd Law,

b) dynamic equilibrium conditions.

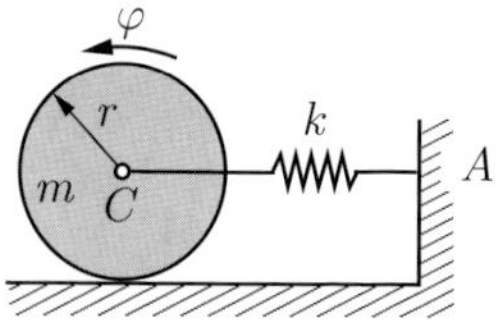

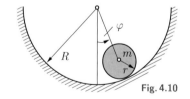

Result: see (**B**) $\ddot{\varphi} + \omega^2 \varphi = 0$, $\omega^2 = \dfrac{2k}{3m}$.

Fig. 4.9

E4.9

Example 4.9 A cylinder (mass m, radius r) rolls without slipping on a circular path (radius R); see Fig. 4.10.

Derive the equation of motion using dynamic equilibrium conditions.

Fig. 4.10

Result: see (**B**) $\ddot{\varphi} + \dfrac{2g}{3(R - r)} \sin \varphi = 0$.

Example 4.10 Two blocks of weights $W_1 = m_1 g$ and $W_2 = m_2 g$ are suspended at a pin-supported rope drum (moment of inertia Θ_A) as shown in Fig. 4.11.

Determine the angular accelerati-on of the drum and the force in rope ① using dynamic equilibrium condi-tions. Neglect the mass of the ropes.

E 4.10

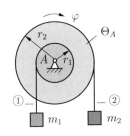

Fig. 4.11

Results: see (**A**)

$$\ddot{\varphi} = \frac{r_2 m_2 - r_1 m_1}{r_1^2 m_1 + r_2^2 m_2 + \Theta_A}\, g\,, \quad S_1 = m_1 g\, \frac{r_2(r_1 + r_2)m_2 + \Theta_A}{r_1^2 m_1 + r_2^2 m_2 + \Theta_A}\,.$$

Example 4.11 An angled arm (mass m) rotates with constant angular ve-locity Ω about point 0 (Fig. 4.12).

Calculate the bending moment, shear force and normal force as func-tions of position using dynamic equi-librium conditions.

E 4.11

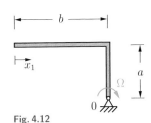

Fig. 4.12

Results: see (**B**) Selected values: $N(x_1) = \mu\Omega^2(bx_1 - x_1^2/2)$,

$$V(x_1) = \mu\Omega^2 a x_1\,, \quad M(x_1) = \mu\Omega^2 a x_1^2/2,\ \mu = m/(a + b)\,.$$

Example 4.12 A wheel (weight $W_1 = m_1 g$, moment of inertia Θ_A) on an inclined plane is connected with a block (weight $W_2 = m_2 g$) by a rope which is guided over an ideal pulley (Fig. 4.13). The wheel rolls on the plane without slipping.

E 4.12

Determine the acceleration of the block applying d'Alembert's principle. Neglect the masses of the rope and the pulley.

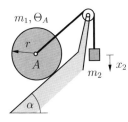

Result: see (**A**) $\ddot{x}_2 = g\, \dfrac{m_2 - m_1 \sin\alpha}{m_1 + m_2 + \dfrac{\Theta_A}{r^2}}\,.$

Fig. 4.13

E4.13

Example 4.13 Two drums are connected by an unwinding rope and carry blocks of weights $m_1 g$ and $m_2 g$ (Fig. 4.14). Drum ① is driven by the moment M_0.

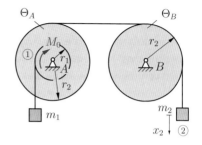

Determine the acceleration of block ② using d'Alembert's principle. Neglect the mass of the ropes.

Fig. 4.14

Result: see (**A**) $\ddot{x}_2 = g \dfrac{1 - \dfrac{m_1 r_1}{m_2 r_2} + \dfrac{M_0}{r_2 m_2 g}}{1 + \dfrac{m_1}{m_2}\left(\dfrac{r_1}{r_2}\right)^2 + \dfrac{\Theta_A}{m_2 r_2^2} + \dfrac{\Theta_B}{m_2 r_2^2}}$.

E4.14

Example 4.14 The system shown in Fig. 4.15 consists of a block (mass M), a homogeneous disk (mass m, radius r) and two springs (spring constant k). The block moves on a frictionless surface; the disk rolls without slipping on the block. A force $F(t)$ acts on the block.

Derive the equations of motion using Lagrange's formalism.

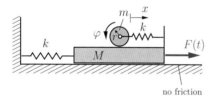

no friction   Fig. 4.15

Result: see (**B**)

$$(M + m)\ddot{x} - mr\ddot{\varphi} + kx = F(t), \quad -m\ddot{x} + \frac{3}{2}mr\ddot{\varphi} + kr\varphi = 0.$$

Example 4.15 Fig. 4.16 shows two blocks of masses m_1 and m_2 which can glide on a friction-less surface. They are coupled by springs (stiffnesses k_1, k_2, k_3).

E4.15

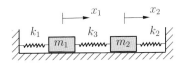

Derive the equations of moti-on using the Lagrange formalism.

Fig. 4.16

Result: see (**A**)
$$m_1\ddot{x}_1 + (k_1 + k_3)x_1 - k_3 x_2 = 0\,, \quad m_2\ddot{x}_2 + (k_2 + k_3)x_2 - k_3 x_1 = 0\,.$$

Example 4.16 Two simple pen-dulums (each mass m, length l) are connected by a spring (spring constant k, unstretched length b) as shown in Fig. 4.17.

E4.16

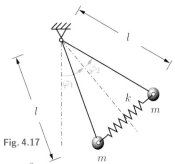

Derive the equations of mo-tion using the Lagrange forma-lism.

Fig. 4.17

Result: see (**B**) $l\ddot{\varphi}_1 + g\sin\varphi_1\cos\varphi_2 = 0\,,$

$$ml\ddot{\varphi}_2 + mg\cos\varphi_1\sin\varphi_2 + k(2l\sin\varphi_2 - b)\cos\varphi_2 = 0\,.$$

Example 4.17   A disk (weight $m_2 g$, moment of inertia Θ_2) glides along a frictionless homogeneous bar of weight $m_1 g$ (Fig. 4.18).

E4.17

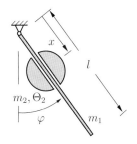

Find the equations of motion using the Lagrange formalism.

Fig. 4.18

Result: see (**A**) $\ddot{x} - x\dot{\varphi}^2 - g\cos\varphi = 0\,,$

$$\left(\frac{m_1 l^2}{3} + m_2 x^2 + \Theta_2\right)\ddot{\varphi} + 2m_2 x\dot{x}\dot{\varphi} + \left(m_1\frac{l}{2} + m_2 x\right)g\sin\varphi = 0\,.$$

E4.18

Example 4.18   A thin half-cylindrical shell of weight $W = mg$ rolls
without sliding on a flat surface
(Fig. 4.19).

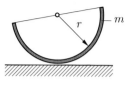

Derive the equation of motion
using the Lagrange formalism.

Fig. 4.19

Result: see **(A)** $\ddot{\varphi}(\pi - 2\cos\varphi) + \dot{\varphi}^2\sin\varphi + \dfrac{g}{r}\sin\varphi = 0.$

E4.19

Example 4.19   A block (mass m_1)
can move horizontally on a smooth
surface (Fig. 4.20). A simple pendulum (mass m_2) is connected to the
block by a pin.

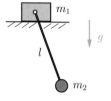

Find the equations of motion
using the Lagrange formalism.

Fig. 4.20

Result: see **(A)** $\ddot{x}\cos\varphi + l\ddot{\varphi} + g\sin\varphi = 0$,

$$(m_1 + m_2)\ddot{x} + m_2l\ddot{\varphi}\cos\varphi - m_2l\dot{\varphi}^2\sin\varphi = 0.$$

4.5 Summary

- By using d'Alembert's inertial forces (pseudo force $\boldsymbol{F}_I = -m\,\boldsymbol{a}$, pseudo moment $M_{IC} = -\Theta_C\,\ddot{\varphi}$), motion can be described by (dynamic) equilibrium conditions. They are given, for example for the plane motion of a rigid body, by

$$F_x + F_{Ix} = 0\,, \qquad F_y + F_{Iy} = 0\,, \qquad M_C + M_{IC} = 0\,.$$

- D'Alembert's principle: the motion of a point mass or a rigid body takes place such that for a virtual displacement the sum of the virtual works done by the applied forces and the inertial forces vanishes:

$$\delta U + \delta U_I = 0\,.$$

Note: constraint forces (reaction forces) do no work!

- The equations of motion of a system with f degrees of freedom can be derived by using the Lagrange equations of the 2nd kind. For conservative systems they are given by

$$\frac{\mathrm{d}}{\mathrm{d}t}\left(\frac{\partial L}{\partial \dot{q}_j}\right) - \frac{\partial L}{\partial q_j} = 0\,, \quad j = 1,\ldots,f\,,$$

$L = T - V$ Lagrangian,

q_j generalized coordinates.

Chapter 5

Vibrations

5

5 Vibrations

———— Objectives: Vibrations play an important role in nature and in engineering. In this chapter we will investigate the behaviour of systems with one or two degrees of freedom which exhibit vibrations. We will restrict ourselves to systems where the equations of motion are linear differential equations. This will already enable us to describe various important features of vibrations. You will learn how to analyse free and forced vibrations with or without damping.

5.1 Basic Concepts

In nature and in engineering, certain quantities, e.g. the position $x(t)$ of a particle, undergo more or less regular changes. Such processes are called *vibrations* or *oscillations*. Examples are the waves of the oceans, the movement of a piston in an engine, or the vibrations in an electrical circuit. Similar processes appear in many areas of our environment. In the following, an introduction to the theory of vibrations of mechanical systems will be given.

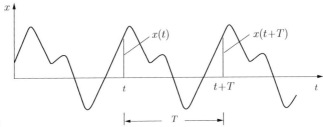

Fig. 5.1

Frequently, a quantity $x(t)$ repeats itself during a motion after a definite time interval T (Fig. 5.1):

$$x(t + T) = x(t). \tag{5.1}$$

In this case the motion is called a *periodic vibration*. The time T is referred to as the *period* of the vibration. The quantity

$$f = \frac{1}{T} \tag{5.2}$$

is the *frequency* of the vibration. It represents the number of cycles per unit of time, where a cycle is the motion completed during a period. The dimension of the frequency is 1/time. Its unit is named after Heinrich Hertz (1857-1894); it is abbreviated as Hz: $1\text{Hz} = 1/\text{s}$.

An important special case of periodic vibrations are *harmonic vibrations*. Here, the behaviour of x in time is given by a cosine or sine function:

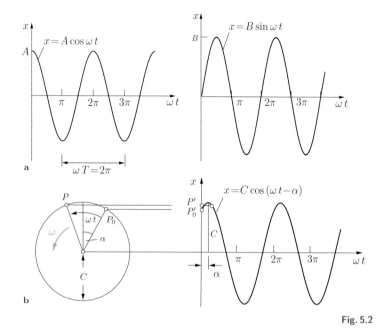

Fig. 5.2

$$x\,(t) = A\cos\omega t \quad \text{or} \quad x\,(t) = B\sin\omega t \tag{5.3}$$

where ω is the *circular frequency* and A or B, respectively, is the *amplitude* of the vibration. Since $\omega T = 2\,\pi$ (see Fig. 5.2a) and $f = 1/T$ we have the following relation between the circular frequency ω and the frequency f:

$$\omega = \frac{2\,\pi}{T} = 2\,\pi\,f\,. \tag{5.4}$$

A vibration that is represented by a pure cosine or sine function is subject to special *initial conditions*. In the case of $x\,(t) = A\cos\omega t$ we have the initial conditions: $x(0) = A$, $\dot{x}(0) = 0$. Similarly, for $x(t) = B\sin\omega t$ the initial conditions are $x(0) = 0$, $\dot{x}(0) = B\omega$. Harmonic vibrations with *arbitrary* initial conditions can always be represented by

$$x\,(t) = C\cos\,(\omega t - \alpha) \tag{5.5}$$

where C is the amplitude and α is referred to as the *phase angle* (see Fig. 5.2b).

Harmonic vibrations (5.5) can also be obtained through a superposition of two vibrations of the form of (5.3). Using the trigonometric formula

$$x\left(t\right) = C\cos\left(\omega t - \alpha\right) = C\cos\omega t\cos\alpha + C\sin\omega t\sin\alpha \qquad (5.6)$$

and the abbreviations

$$A = C\cos\alpha, \quad B = C\sin\alpha \qquad (5.7)$$

we get

$$x\left(t\right) = A\cos\omega t + B\sin\omega t\,. \qquad (5.8)$$

Thus, the representations (5.5) and (5.8) are equivalent; they are interchangeable. If C and α are given, we obtain A and B from (5.7). On the other hand, if A and B are given, (5.7) can be solved for C and α:

$$C = \sqrt{A^2 + B^2}, \quad \alpha = \arctan\frac{B}{A}\,. \qquad (5.9)$$

A harmonic oscillation can be generated by a point P (initial position P_0) which moves on a circular path (radius C) with constant angular velocity ω (Fig. 5.2b). Then the projection P' on a vertical straight line (or on any diameter) performs a harmonic vibration. This is shown in Fig. 5.2b.

A vibration with a constant amplitude is called an *undamped vibration*. If the amplitude decreases with increasing time (Fig. 5.3a), the vibration is referred to as *damped*. In the case of an increasing amplitude one speaks of an unstable vibration (Fig. 5.3b).

Vibrations may be classified according to various criteria. For example, one may classify vibration by the number of degrees of freedom $(1, 2, \ldots, n)$ involved. For greatest transparency, we will restrict our presentation to systems with one or two degrees of freedom. This will already enable us to describe various important features of the vibrations without unneeded complications.

One may also classify vibrations according to the type of the differential equation that describes the motion of the system. Thus, in the case of a linear (nonlinear) differential equation one speaks of *linear (nonlinear)* vibrations.

A third classification is based on the mechanism that generates the vibrations. We will restrict ourselves to two cases: free vibrations and forced vibrations. *Free vibrations* occur in a system that is acted upon only by forces within the system (e.g., weights, forces in springs); there are no external forces. *Forced vibrations* are generated under the influence of external forces.

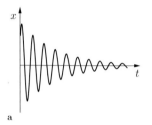
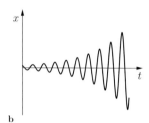

a   b   **Fig. 5.3**

5.2 Free Vibrations

In the following sections, the behaviour of *linear* systems with *one* degree of freedom will be investigated. The vibrations will be free but may be damped or undamped.

5.2.1 Undamped Free Vibrations

At first we restrict ourselves to the investigation of undamped vibrations. As an example let us consider a block (mass m) that moves on a smooth surface (Fig. 5.4a). It is connected to a wall with a linear spring (spring constant k). To derive the equation of motion, we introduce the coordinate x as shown in Fig. 5.4b: $x = 0$ is the equilibrium position of the block (unstressed spring). The only force in the horizontal direction is the force $k\,x$ in the spring. It is a restoring force, i.e., it acts in the direction opposite to the displacement from the equilibrium position. Thus, Newton's law (1.38) yields

$$\rightarrow:\quad m\ddot{x} = -\,k\,x \quad\rightarrow\quad m\ddot{x} + k\,x = 0. \tag{5.10}$$

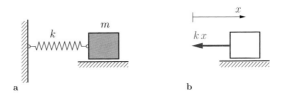

Fig. 5.4

With the abbreviation

$$\omega^2 = \frac{k}{m} \tag{5.11}$$

we obtain

$$\ddot{x} + \omega^2 x = 0 . \tag{5.12}$$

This is a linear homogeneous differential equation of second order; it has constant coefficients. The general solution of (5.12) is given by

$$x(t) = A \cos \omega t + B \sin \omega t \tag{5.13}$$

where A and B are constants of integration. They can be calculated from the given initial conditions $x(0) = x_0$ and $\dot{x}(0) = v_0$ which yields

$$A = x_0 \quad \text{and} \quad B = \frac{v_0}{\omega} . \tag{5.14}$$

Thus, (5.13) becomes

$$x(t) = x_0 \cos \omega t + \frac{v_0}{\omega} \sin \omega t . \tag{5.15}$$

According to Section 5.1, the general solution (5.13) is equivalent to

$$x(t) = C \cos(\omega t - \alpha) \tag{5.16}$$

where now C and α are the constants of integration. They can also be calculated from the initial conditions, however, they follow immediately from (5.9) and (5.14):

$$C = \sqrt{x_0^2 + (v_0/\omega)^2}, \quad \alpha = \arctan \frac{v_0}{\omega x_0}. \tag{5.17}$$

According to (5.16) the block performs a harmonic vibration with circular frequency $\omega = \sqrt{k/m}$. The circular frequency of a free vibration is also called the *natural frequency* or the *eigenfrequency*.

Let us now consider a block (mass m) that is suspended by a linear spring (spring constant k), see Fig. 5.5a. We assume that the block performs *vertical* oscillations. When the block is attached to the spring, the spring undergoes an elongation $x_{st} = mg/k$ from its unstressed (natural) length due to the weight $W = mg$ of the block. We measure the location x of the block from this equilibrium position (Fig. 5.5b). The forces that act in the vertical direction on the displaced block are the weight W and the spring force (restoring force) $F_k = k(x_{st} + x)$. Newton's law (1.38) leads to

$$\downarrow: \quad m\ddot{x} = mg - k(x_{st} + x) \quad \rightarrow \quad m\ddot{x} + kx = 0.$$

This is again the equation of motion (5.10). Note, the weight of the block has no influence on the vibration of the system consisting of a mass and a spring. We therefore do not have to consider the weight of such systems if we measure the displacement x from the position of static equilibrium.

The natural frequency of the vertical vibration of a system with one degree of freedom may be calculated from the static displacement due to its weight. Then the mass of the system and the spring constant need not be known. For example, the spring in Fig. 5.5a undergoes an elongation $x_{st} = mg/k$ due to the weight $W = mg$ of the block. Comparison of $k x_{st} = mg$ with (5.11) yields

$$\omega^2 = g/x_{st}. \tag{5.18}$$

The motions of many mechanical systems are described by differential equations of the type (5.12). These systems perform har-

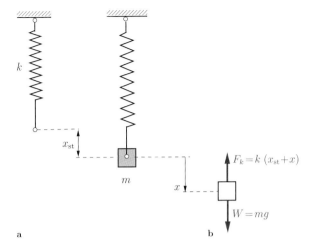

Fig. 5.5   a   b

monic vibrations. For example, the motion of a *simple pendulum* (also called an *ideal pendulum*) as shown in Fig. 5.6a is governed by the differential equation

$$\ddot{\varphi} + \frac{g}{l}\sin\varphi = 0\,; \tag{5.19}$$

see Section 1.2.6. If we assume that the displacements are small $(\sin\varphi \approx \varphi)$, we obtain the differential equation $\ddot{\varphi} + (g/l)\,\varphi = 0$ for a harmonic vibration. The natural frequency of a simple pendulum is thus given by

$$\omega = \sqrt{g/l}\,. \tag{5.20}$$

As a further example let us consider a rigid body which is suspended at a fixed point (pin support A) and undergoes vibratory motion (Fig. 5.6b). Such a system is called a *compound pendulum*. The center of gravity C is located at a distance l from pin A. To derive the equation of motion we apply the principle of angular momentum (3.33). With the coordinate φ as shown in Fig. 5.6b (positive counterclockwise), the moment $M_A = -mgl\sin\varphi$ of the weight W and with the moment of inertia Θ_A we obtain

$$\overset{\curvearrowleft}{A}:\quad \Theta_A\,\ddot{\varphi} = -\,mgl\sin\varphi \quad \rightarrow \quad \Theta_A\,\ddot{\varphi} + mgl\sin\varphi = 0\,.$$

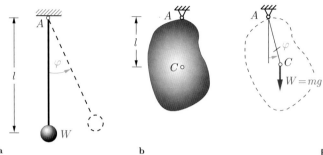

a   b   Fig. 5.6

If we assume small displacements ($\sin \varphi \approx \varphi$), this equation reduces to

$$\ddot{\varphi} + \omega^2 \varphi = 0$$

where $\omega^2 = mgl/\Theta_A$. We now introduce the effective length $l_{\text{eff}} = \Theta_A/(ml) = r_g^2/l$ of the pendulum. Then the natural frequency of the compound pendulum may be written as $\omega = \sqrt{g/l_{\text{eff}}}$ in analogy to (5.20). Thus, a compound pendulum has the same natural frequency as an ideal pendulum with length l_{eff}.

All the systems considered so far are conservative systems. Therefore, the principle of conservation of energy is valid:

$$T + V = T_0 + V_0 = E = \text{const}, \tag{5.21}$$

where E is the total energy of the system. Using the general solution (5.16) of the equation of motion (5.10) and the trigonometric formulae $\sin^2 \beta = \frac{1}{2}(1 - \cos 2\beta)$, $\cos^2 \beta = \frac{1}{2}(1 + \cos 2\beta)$ we can express the kinetic and potential energies as

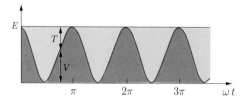

Fig. 5.7

$$T = \tfrac{1}{2}m\dot{x}^2 = \tfrac{1}{2}m\omega^2 C^2 \sin^2(\omega t - \alpha)$$

$$= \tfrac{1}{4}m\omega^2 C^2[1 - \cos(2\omega t - 2\alpha)],$$

$$V = \tfrac{1}{2}kx^2 = \tfrac{1}{2}k C^2 \cos^2(\omega t - \alpha)$$

$$= \tfrac{1}{4}k C^2[1 + \cos(2\omega t - 2\alpha)].$$

(5.22)

Both energies are represented by periodic functions with frequency 2ω. Their amplitudes are equal since $m\omega^2 = k$. The energies are depicted in Fig. 5.7. Note that there is a periodic exchange between the potential energy and the kinetic energy. When the kinetic (potential) energy is zero, the potential (kinetic) energy attains its maximum value. The sum of both energies is constant, namely equal to the total energy E (conservation of energy).

Example 5.1 A rod (length l, with negligible mass) carries a mass m at its upper end. It is supported by a linear spring (spring constant k), see Fig. 5.8a.

Describe the motion of the rod if it is displaced from its vertical position (small displacement) and then released (no initial velocity).

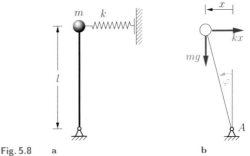

Fig. 5.8   **a**   **b**

Solution We introduce the coordinates φ (positive counterclockwise) and x as shown in Fig. 5.8b. Then the elongation of the spring is given by $x = l \sin \varphi$. The principle of angular momentum (1.67) leads to the equation of motion ($\Theta_A = ml^2$):

$\overset{\curvearrowleft}{A}$: $ml^2\ddot{\varphi} = l\sin\varphi\,mg - l\cos\varphi\,k\,x$

$\rightarrow$ $ml\ddot{\varphi} - mg\sin\varphi + k\,l\sin\varphi\cos\varphi = 0$.

We assume that the displacements are small. Then with $\sin\varphi \approx \varphi$, $\cos\varphi \approx 1$ we obtain

$$ml\ddot{\varphi} - mg\varphi + k\,l\varphi = 0 \quad \rightarrow \quad \ddot{\varphi} + \frac{k\,l - mg}{ml}\varphi = 0 \,.$$

This is the differential equation of a harmonic oscillator if $k\,l > mg$. Comparison with (5.12) yields the eigenfrequency:

$$\omega^2 = \frac{k\,l - mg}{ml} \quad \rightarrow \quad \underline{\underline{\omega = \sqrt{\frac{k\,l - mg}{ml}}}}\,.$$

The constants of integration in the general solution $\varphi(t) = A\cos\omega t + B\sin\omega t$ can be calculated from the initial conditions $\varphi(0) = \varphi_0$ and $\dot{\varphi}(0) = 0$. We obtain $A = \varphi_0$ and $B = 0$. Thus, the motion of the rod is described by

$$\underline{\underline{\varphi(t) = \varphi_0\cos\omega t}}\,.$$

If $k\,l < mg$, the restoring moment due to the spring force is smaller than the moment due to the weight: the rod falls. In the special case $k\,l = mg$, the eigenfrequency is zero: the rod is in (static) equilibrium in the displaced position.

5.2.2 Spring Constants of Elastic Systems

The relation between the spring force F and the elongation Δl of a linear spring is given by $F = k\,\Delta l$. The spring constant k is therefore characterized by

$$k = \frac{F}{\Delta l}\,. \tag{5.23}$$

In many systems with elastic components there also exists a linear relation between the force and the deformation. Let us first consider a massless bar (length l, axial rigidity EA) which carries a mass m at one end (Fig. 5.9a). If the mass is displaced downward and the bar undergoes an elongation Δl, then a restoring force F acts on the mass. A force of equal magnitude is exerted on the bar (action = reaction). The relation

$$\Delta l = \frac{Fl}{EA}$$

is known from Volume 2. In analogy to (5.23) we obtain the "spring constant" or *stiffness* of an elastic bar:

$$k = \frac{F}{\Delta l} = \frac{EA}{l} .$$ (5.24)

Therefore, the original system (elastic bar with end mass) in Fig. 5.9a is equivalent to a system consisting of a spring and a mass (Fig. 5.9b) and can be replaced by it if the spring constant is chosen according to (5.24). Note that the stiffness of the bar is equal to the force required to produce a unit elongation.

Let us now consider a massless cantilever beam (length l, flexural rigidity EI) with a mass m at its free end as shown in Fig. 5.9c. If the mass is displaced downward a restoring force F acts on it. A force of equal magnitude is exerted on the beam and causes the deflection

$$w = \frac{Fl^3}{3\,EI} ,$$

see Volume 2. Thus, we obtain the spring constant

$$k = \frac{F}{w} = \frac{3\,EI}{l^3}$$ (5.25)

of the elastic beam. If the spring constant of the system in Fig. 5.9b is chosen according to (5.25), then this simple system is equivalent to the massless cantilever beam with a single mass at its free end.

We finally determine the spring constant of a massless shaft (length l, torsional rigidity GI_T) under torsion (Fig. 5.9d). It follows from the linear relation between the angle of twist ϑ and the torque M_T (see Volume 2):

$$\vartheta = \frac{M_T l}{GI_T} \quad \rightarrow \quad k_T = \frac{M_T}{\vartheta} = \frac{GI_T}{l} .$$ (5.26)

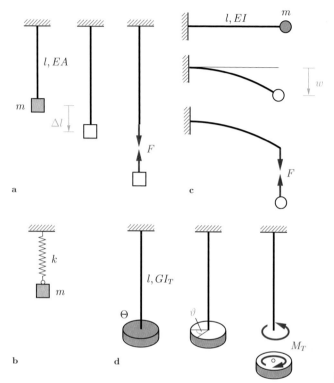

Fig. 5.9

The constant k_T has the dimensions moment/angle. If a disk (moment of inertia Θ) is fixed to the end of the shaft and undergoes torsional vibrations, then the motion is described by $\Theta\ddot{\vartheta}+k_T\vartheta = 0$.

The motion of a mass may cause elongations of *several* springs in a system. Let us first consider the case of two *springs in parallel* (Fig. 5.10a). The two springs (spring constants k_1 and k_2) undergo the same elongation when the mass is displaced. They can be replaced by an equivalent single spring with the spring constant k^*. To determine k^*, we displace the mass by an amount x. This displacement causes the forces $F_1 = k_1\,x$ and $F_2 = k_2\,x$ in the springs, and the mass is acted upon by the force $F = F_1 + F_2$. Since the single spring is assumed to be equivalent to the two springs, the same displacement x of the equivalent system has to

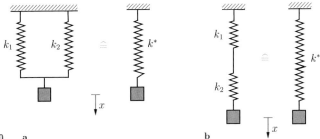

Fig. 5.10   a   b

generate the same force $F = k^* x$. Thus,

$$F = k_1 x + k_2 x = k^* x \quad \rightarrow \quad k^* = k_1 + k_2 .$$

In the case of a system with arbitrarily many parallel springs (spring constants k_j), the spring constant of the equivalent spring is given by the sum of the individual spring constants:

$$k^* = \sum k_j . \tag{5.27}$$

Let us now consider two *springs in series* (Fig. 5.10b). In this case the total elongation x is the sum $x_1 + x_2$ of the elongations of the springs and the same force F acts in each spring. With $F = k_1 x_1 = k_2 x_2$ and $x = x_1 + x_2$ we obtain

$$x = \frac{F}{k_1} + \frac{F}{k_2} = \frac{F}{k^*} \quad \rightarrow \quad \frac{1}{k^*} = \frac{1}{k_1} + \frac{1}{k_2} .$$

In the case of arbitrarily many springs in series, the spring constant of the equivalent spring is found from

$$\frac{1}{k^*} = \sum \frac{1}{k_j} . \tag{5.28}$$

We now introduce the *flexibility (compliance)* $f = 1/k$ of a spring (not to be confused with the frequency). According to (5.28) the flexibility of arbitrarily many springs in series is given by the sum of the individual flexibilities:

$$f^* = \sum f_j \, . \tag{5.29}$$

Note that the flexibility is the elongation of the spring due to a unit force.

E5.2   **Example 5.2** An elastic beam (flexural rigidity EI) with negligible mass supports a box (mass m) as shown in Fig. 5.11a.
Find the natural frequency of the system.

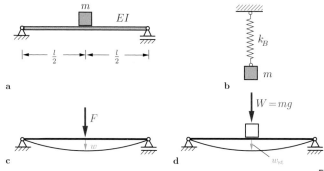

Fig. 5.11

Solution We reduce the system consisting of the massless beam and the mass to the equivalent simple system of a spring and a mass (Fig. 5.11b). To determine the equivalent spring constant k_B we subject the beam to a force F which acts at the location of the box (Fig. 5.11c). This force produces the deflection (see Volume 2)

$$w = \frac{Fl^3}{48\,EI} \, . \tag{a}$$

Thus, in analogy to (5.25) we obtain

$$k_B = \frac{F}{w} = \frac{48\,EI}{l^3} \, ,$$

and (5.11) yields the natural frequency

$$\underline{\underline{\omega = \sqrt{\frac{k_B}{m}} = \sqrt{\frac{48\,EI}{ml^3}}}} \, .$$

According to (5.18) it is also possible to determine the natural frequency with the aid of the static deflection caused by the weight $W = mg$ of the mass m. This deflection is given by (Fig. 5.11d)

$$w_{st} = \frac{Wl^3}{48\,EI} = \frac{mgl^3}{48\,EI} \, .$$

Substitution into (5.18) again yields

$$\omega = \sqrt{\frac{g}{w_{st}}} = \sqrt{\frac{48\,EI}{ml^3}} \, .$$

Example 5.3 The systems in the Figs. 5.12a,b consist of massless beams (flexural rigidity EI), a spring (spring constant k) and a box (mass m).

    Determine the natural frequencies of the systems.

E5.3

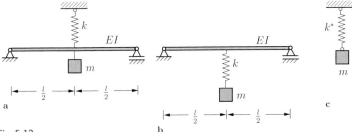

Fig. 5.12

Solution We replace both systems with the simple equivalent system shown in Fig. 5.12c.

a) If the mass in Fig. 5.12a is displaced, the deflection at the middle of the beam and the elongation of the spring are equal. Therefore, the beam and the spring act as springs in parallel. The spring constant of the beam can be taken from Example 5.2:

$$k_B = \frac{48\,EI}{l^3} \, .$$

Thus, Equation (5.27) yields the spring constant k^* of the equivalent spring,

$$k^* = k + k_B = k + \frac{48\,EI}{l^3},$$

and the natural frequency of the system is obtained as

$$\underline{\underline{\omega}} = \sqrt{\frac{k^*}{m}} = \underline{\sqrt{\frac{k\,l^3 + 48\,EI}{ml^3}}}\,.$$

b) We now consider the system in Fig. 5.12b. The displacement of the mass is the sum of the deflection at the middle of the beam and the elongation of the spring. Therefore, the beam and the spring act as springs in series. The spring constant k^* of the equivalent spring now follows from (5.28):

$$\frac{1}{k^*} = \frac{1}{k} + \frac{1}{k_B} \quad \rightarrow \quad k^* = \frac{k\,k_B}{k + k_B}.$$

This yields the natural frequency

$$\underline{\underline{\omega}} = \sqrt{\frac{k^*}{m}} = \underline{\sqrt{\frac{48\,k\,EI}{(k\,l^3 + 48\,EI)\,m}}}\,.$$

The natural frequency of system b) is smaller than the one of system a): the equivalent spring of system b) is softer than the one of system a).

Example 5.4 The frame in Fig. 5.13a consists of two elastic columns ($h = 3\,\mathrm{m}$, $E = 210\,\mathrm{GPa}$, $I = 3500\,\mathrm{cm}^4$) and a rigid horizontal beam. It carries a box ($m = 10^5\,\mathrm{kg}$).

Calculate the natural frequency of the system. Neglect the masses of the members of the frame.

Solution Fig. 5.13b shows the deformed frame. Since the columns are elastic and the beam is rigid, the box is displaced horizontally. The vibrations of the system can be described by the equivalent simple system in Fig. 5.13c. As the system is symmetrical, we will now determine the spring constant k_c of the equivalent spring for a single column. Consider the column as shown in Fig. 5.13d. The rigid beam at the upper end of the column forces a parallel motion (see Volume 1, Section 5.1.1). If the column is subjected to a force F at point A, this point undergoes a deflection w. The spring constant k_c of the column follows from $k_c = F/w$.

The system in Fig. 5.13d is statically indeterminate of first degree. We choose the moment at the top to be the statical redundant X and remove the support (Fig. 5.13e). Then the displacement w and the angle φ at the top end are obtained as

$$w = \frac{Fh^3}{3\,EI} - \frac{Xh^2}{2\,EI}, \qquad \varphi = \frac{Fh^2}{2\,EI} - \frac{Xh}{EI},$$

see Volume 2, Table 4.3. The condition $\varphi = 0$ yields the support moment: $X = Fh/2$. Thus,

$$w = \frac{Fh^3}{3\,EI} - \frac{Fh^3}{4\,EI} = \frac{Fh^3}{12\,EI},$$

and the spring constant of one column follows as

$$k_c = \frac{F}{w} = \frac{12\,EI}{h^3}.$$

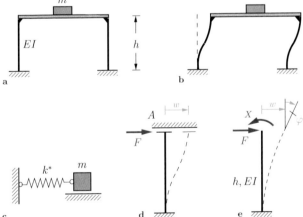

Fig. 5.13

The frame has two columns in parallel. Therefore, the spring constant of the frame is given by $k^* = 2\,k_c$. Hence, the natural frequency of the system is found to be

$$\underline{\underline{\omega}} = \sqrt{\frac{k^*}{m}} = \sqrt{\frac{24\,EI}{mh^3}}$$

which leads to

$$\underline{\underline{\omega = 8.1\ \text{s}^{-1}}} \quad \text{or} \quad f = \frac{\omega}{2\,\pi} = 1.3\ \text{Hz}\,.$$

5.2.3 Damped Free Vibrations

Experience shows that a free vibration with a constant amplitude never occurs in reality. The amplitudes of the vibrations of real systems decay with time and, finally, the vibrations die out. This is due to the friction forces and damping forces which are always present in real systems (for example, friction at the supports, air resistance). The systems lose mechanical energy during the motion (energy dissipation). Therefore, conservation of energy is not valid for damped vibrations.

As an example of damped vibrations let us consider the system shown in Fig. 5.14a. It consists of a block (mass m) which moves on a rough horizontal surface (coefficient of kinetic friction μ). We assume dry friction (Coulomb friction, see Volume 1). The friction force $R = \mu N = \mu mg$ is always oriented opposite to the direction of the velocity. If the block moves to the right (left) the friction force is directed to the left (right), see Fig. 5.14b. We measure the coordinate x from the position of the block when the spring is unstressed. The restoring force is given by $k\,x$. Newton's law (1.38) yields the equation of motion:

$$\rightarrow:\quad m\ddot{x} = \begin{cases} -k\,x - R & \text{for}\quad \dot{x} > 0, \\ -k\,x + R & \text{for}\quad \dot{x} < 0 \end{cases}$$

$$\rightarrow\quad m\ddot{x} + k\,x = \begin{cases} -R & \text{for}\quad \dot{x} > 0, \\ +R & \text{for}\quad \dot{x} < 0\,. \end{cases}$$

If we introduce the definitions

$$\omega^2 = \frac{k}{m}, \quad r = \frac{R}{k},$$

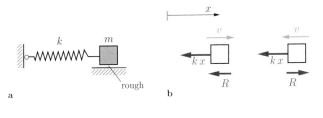

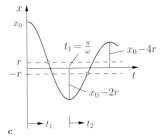

Fig. 5.14

we obtain

$$\ddot{x} + \omega^2 x = \begin{cases} -\omega^2 r & \text{for} \quad \dot{x} > 0, \\ +\omega^2 r & \text{for} \quad \dot{x} < 0. \end{cases} \tag{a}$$

Thus, two different equations describe the motion of the block to the right and to the left, respectively.

We use the variable t_1 during the first part of the vibration and choose the initial conditions $x(t_1 = 0) = x_0 > 0$, $\dot{x}(t_1 = 0) = 0$, see Fig. 5.14c. Then the block first moves to the left: $\dot{x} < 0$. Hence, we use

$$\ddot{x} + \omega^2 x = \omega^2 r. \tag{b}$$

In contrast to the equation of motion (5.12), the right-hand side of this equation is *not* zero. A differential equation of this type is called an *inhomogeneous* differential equation. Its general solution is composed of the solution x_h of the homogeneous differential equation ($\ddot{x} + \omega^2 x = 0$), usually called the *homogeneous solution*, and a particular solution x_p of the inhomogeneous differential equation, yielding the general solution

$$x = x_h + x_p .$$

According to (5.13) the solution x_h is given by

$$x_h (t_1) = A_1 \cos \omega t_1 + B_1 \sin \omega t_1.$$

The particular solution

$$x_p = r$$

can be found by inspection of the inhomogeneous differential equation. Thus,

$$x (t_1) = A_1 \cos \omega t_1 + B_1 \sin \omega t_1 + r .$$

The constants of integration A_1 and B_1 follow from the initial conditions:

$$x (t_1 = 0) = A_1 + r = x_0 \quad \rightarrow \quad A_1 = x_0 - r,$$
$$\dot{x} (t_1 = 0) = \omega B_1 = 0 \quad \rightarrow \quad B_1 = 0 .$$

Hence, the motion to the left during the first half cycle of the vibration is described by

$$x (t_1) = (x_0 - r) \cos \omega t_1 + r ,$$

$$\dot{x} (t_1) = - (x_0 - r) \omega \sin \omega t_1. \tag{c}$$

At time $t_1 = \pi/\omega$, the displacement is $x(\pi/\omega) = -x_0 + 2\,r$ and the velocity vanishes: $\dot{x}(\pi/\omega) = 0$. Then the motion changes its direction: the block moves to the right and we have to use the equation of motion (see (a))

$$\ddot{x} + \omega^2 x = - \omega^2 r . \tag{d}$$

We will use the variable t_2 to describe the motion during the second half cycle of the vibration (Fig. 5.14c). The general solution of (d) is given by

$$x (t_2) = A_2 \cos \omega t_2 + B_2 \sin \omega t_2 - r .$$

The displacement and the velocity at the beginning of the second half cycle have to coincide with the corresponding quantities at the end of the first half cycle. Therefore, the constants of integration A_2 and B_2 can be calculated from the matching conditions

$$x\left(t_2 = 0\right) = x\left(t_1 = \frac{\pi}{\omega}\right) \quad \rightarrow \quad A_2 = -x_0 + 3\,r\,,$$

$$\dot{x}\left(t_2 = 0\right) = \dot{x}\left(t_1 = \frac{\pi}{\omega}\right) \quad \rightarrow \quad B_2 = 0\,.$$

The motion during the second half cycle is therefore described by

$$x\left(t_2\right) = -\left(x_0 - 3\,r\right)\cos\omega t_2 - r\,. \tag{e}$$

The displacement-time graph of the vibration is shown in Fig. 5.14c. The first of the equations (c) represents half a cosine which is shifted by $+r$ in the x-direction and has the amplitude $x_0 - r$. The function (e) is shifted by $-r$ and has the amplitude $x_0 - 3\,r$.

The following cycles of the vibration can be determined in an analogous manner. The amplitudes are decreased by $2r$ after each change of direction of the motion. If the displacement of the block is smaller than r at a position where the velocity is zero, then the restoring force in the spring is too small to overcome the static friction force and the block comes to a rest.

Resisting forces caused by friction in fluids were already introduced in Section 1.2.4. Such forces may, for example, be generated in the shock absorber of a car. We will restrict ourselves to the case of a linear relation between the velocity v and the resisting force F_d:

$$F_d = d\,v\,.$$

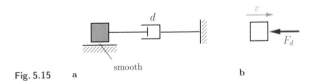

Fig. 5.15   a   smooth   b

The quantity d is called the *coefficient of viscous damping* or the *damping coefficient*; it has the dimensions force/velocity. We will depict a damper (or dashpot) symbolically as shown in Fig. 5.15a. The resisting force that is exerted upon a body in motion is directed opposite to its velocity (Fig. 5.15b).

Let us consider the system consisting of a block, a spring and a dashpot as shown in Fig. 5.16a. We measure the coordinate x from the equilibrium position. In this case we need not consider the weight of the block. With the restoring force $k\,x$ of the spring and the viscous damping force $d\,\dot{x}$ in the dashpot (Fig. 5.16b) we obtain the equation of motion

$$\downarrow:\quad m\ddot{x} = -\,k\,x - d\,\dot{x} \quad \rightarrow \quad m\ddot{x} + d\,\dot{x} + k\,x = 0\,. \tag{5.30}$$

We now introduce the abbreviations

$$2\,\xi = \frac{d}{m}\,,\quad \omega^2 = \frac{k}{m}\,. \tag{5.31}$$

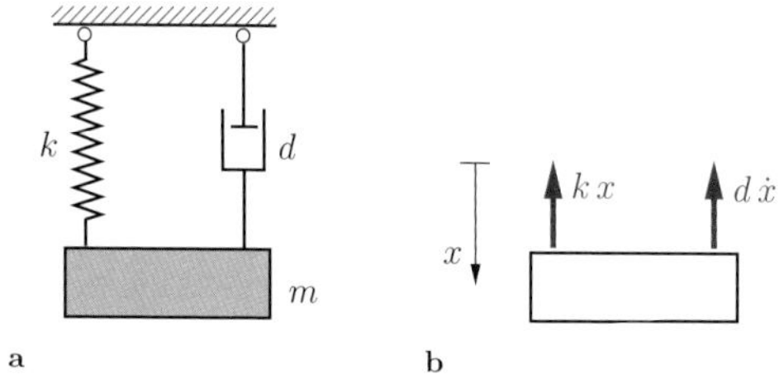

Fig. 5.16

The constant ξ is called the *normalized damping coefficient* and ω is the natural frequency of the undamped vibrations (see (5.11)). This yields the differential equation of the damped vibrations:

$$\ddot{x} + 2\,\xi\dot{x} + \omega^2\,x = 0\,. \tag{5.32}$$

We want to find the general solution of this differential equation. Since it has constant coefficients we assume a solution of the form

$$x = A\,\mathrm{e}^{\lambda t} \tag{5.33}$$

where e is the base of the natural logarithm. The constants A and λ are as yet unknown. In order to determine λ, we introduce

(5.33) into (5.32) and obtain the *characteristic equation*

$$\lambda^2 + 2\xi\lambda + \omega^2 = 0 \,. \tag{5.34}$$

This quadratic equation for λ has the two solutions

$$\lambda_{1,2} = -\xi \pm \sqrt{\xi^2 - \omega^2} \,. \tag{5.35}$$

If we introduce the *damping ratio* (Ernst Lehr, 1896-1944)

$$\zeta = \frac{\xi}{\omega} = \frac{d}{2\sqrt{m\,k}} \tag{5.36}$$

we can write (5.35) in the form

$$\lambda_{1,2} = -\xi \pm \omega\sqrt{\zeta^2 - 1} \,. \tag{5.37}$$

Depending on the value of ζ, the solutions of (5.32) exhibit different behaviours. We distinguish between three cases.

1. Overdamped System: $\zeta > 1$

In this case, both solutions λ_1 and λ_2 of (5.37) are real numbers: $\lambda_{1,2} = -\xi \pm \mu$, where $\mu = \omega\sqrt{\zeta^2 - 1}$. Each λ_i is associated with a solution of the differential equation (5.32). The general solution of (5.32) is represented by a linear combination of both solutions:

$$x\,(t) = A_1\,e^{\lambda_1 t} + A_2\,e^{\lambda_2 t} = e^{-\xi t}(A_1\,e^{\mu t} + A_2\,e^{-\mu t}) \,. \tag{5.38}$$

The constants A_1 and A_2 can be calculated from the initial conditions $x\,(0) = x_0$ and $\dot{x}\,(0) = v_0$. Since $\xi > \mu$, Equation (5.38) represents a motion which decays exponentially. Since there is no period, the motion is called *aperiodic*. The displacement has at most one maximum and at most one zero value. This motion is not a vibration: the block *slides* back to its equilibrium position without oscillating. Fig. 5.17 shows several graphs of equation (5.38) for different initial conditions.

2. Critically Damped System: $\zeta = 1$

If $\zeta = 1$, the radical in (5.37) is equal to zero and the characteristic equation (5.34) has repeated roots: $\lambda_1 = \lambda_2 = -\xi$, see (5.37).

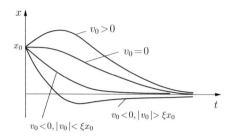

Fig. 5.17

The general solution of (5.32) is then given by

$$x(t) = A_1 e^{\lambda_1 t} + A_2 t e^{\lambda_1 t} = (A_1 + A_2 t) e^{-\xi t}. \tag{5.39}$$

This also represents a motion which decays exponentially. As in the case of an overdamped system, the decay occurs without oscillation. It can be shown that for $\zeta = 1$ the displacement converges to zero faster than in the case of an overdamped system ($\zeta > 1$). Since the value $\zeta = 1$ separates the aperiodic motions from the oscillatory motions, this case is referred to as *critical damping*. Critical damping is made use of in the design of measuring devices.

For $\zeta = 1$ we have $\xi = \omega$ according to (5.36). Then, using (5.31), we obtain the *critical damping coefficient*

$$d_c = 2\sqrt{mk}.$$

Note that the damping factor ζ according to (5.36) may also be written as the ratio of the damping coefficient d to the critical damping coefficient d_c:

$$\zeta = \frac{d}{d_c}.$$

3. Underdamped System: $\zeta < 1$

In the case of an underdamped system ($\zeta < 1$) the radical in (5.37) is negative. Therefore we write the two solutions of the characteristic equation in the form

$$\lambda_{1,2} = -\xi \pm i\omega\sqrt{1-\zeta^2} = -\xi \pm i\omega_d, \quad (i = \sqrt{-1})$$

with

$$\omega_d = \omega\sqrt{1-\zeta^2}\,. \tag{5.40}$$

This yields the general solution of the differential equation (5.32):

$$x(t) = A_1 e^{\lambda_1 t} + A_2 e^{\lambda_2 t} = e^{-\xi t}(A_1 e^{i\omega_d t} + A_2 e^{-i\omega_d t})\,.$$

Using $e^{\pm i\omega_d t} = \cos\omega_d t \pm i\sin\omega_d t$ we obtain

$$\begin{aligned}x(t) &= e^{-\xi t}[(A_1 + A_2)\cos\omega_d t + i(A_1 - A_2)\sin\omega_d t]\\ &= e^{-\xi t}(A\cos\omega_d t + B\sin\omega_d t),\end{aligned}$$

where we have introduced two new real constants A and B. According to Section 5.1 the displacement $x(t)$ can also be written in the form

$$x(t) = C e^{-\xi t}\cos(\omega_d t - \alpha)\,. \tag{5.41}$$

Thus, the motion of an underdamped system is a vibration with exponentially decaying amplitudes. The constants of integration C and α can be calculated from the initial conditions. As $t \to \infty$,

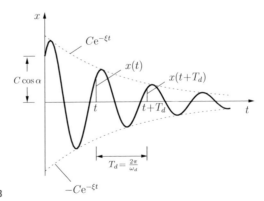

Fig. 5.18

the displacement converges to zero. Fig. 5.18 shows the graph of equation (5.41) and the envelopes $\pm C\,e^{-\xi t}$.

The circular frequency ω_d of the damped vibration is smaller than the circular frequency ω of the undamped vibration (see (5.40)). Therefore, the period $T_d = 2\,\pi/\omega_d$ is larger than the period T of the corresponding undamped vibration.

The displacements at time t and time $t + T_d$ are given by

$$x\,(t) = C\,e^{-\xi t}\cos\,(\omega_d\,t - \alpha)$$

and

$$\begin{aligned} x\,(t + T_d) &= C\,e^{-\xi(t+T_d)}\cos\,[\omega_d(t + T_d) - \alpha] \\ &= C\,e^{-\xi(t+T_d)}\cos\,(\omega_d\,t - \alpha)\,, \end{aligned}$$

respectively. The ratio of the two displacements is therefore given by

$$\frac{x\,(t)}{x\,(t + T_d)} = e^{\xi T_d}\,. \tag{5.42}$$

The logarithm

$$\delta = \ln\frac{x\,(t)}{x\,(t + T_d)} = \xi T_d = \frac{2\,\pi\xi}{\omega_d} = 2\,\pi\frac{\zeta}{\sqrt{1 - \zeta^2}} \tag{5.43}$$

of this ratio is referred to as the *logarithmic decrement*. If the logarithmic decrement δ can be determined from experiments, the damping ratio ζ can be calculated according to (5.43).

E5.5

Example 5.5   Fig. 5.19a shows a system which consists of a rigid bar (with negligible mass), a spring, a dashpot and a point mass. Find the condition which the damping coefficient d has to satisfy in order for the system to be underdamped. Determine the solution of the equation of motion for the initial conditions $\varphi\,(0) = 0$, $\dot{\varphi}\,(0) = \dot{\varphi}_0$. Assume small displacements.

Solution   We introduce the coordinate φ as shown in Fig. 5.19b. The moment of inertia with respect to point A is given by $\Theta_A = (2\,a)^2 m$, the restoring force of the spring is $F_k = k\,a\,\varphi$ and the viscous damping force is $F_d = d\,(3\,a)\dot{\varphi}$. The principle of angular momentum yields the equation of motion:

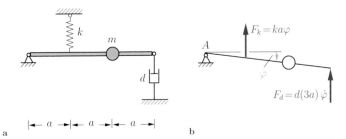

Fig. 5.19

$$\widehat{A}^{\curvearrowright}: \quad \Theta_A \ddot{\varphi} = -a\,F_k - 3\,a\,F_d \quad \rightarrow \quad 4\,m\ddot{\varphi} + 9\,d\,\dot{\varphi} + k\,\varphi = 0\,.$$

We introduce the abbreviations $2\,\xi = 9\,d/(4m)$ and $\omega^2 = k/(4m)$ and obtain

$$\ddot{\varphi} + 2\,\xi\,\dot{\varphi} + \omega^2\varphi = 0\,,$$

cf. (5.32). The system is underdamped if $\zeta < 1$:

$$\zeta = \frac{\xi}{\omega} = \frac{9\,d}{8\,m}2\sqrt{\frac{m}{k}} = \frac{9\,d}{4\sqrt{mk}} < 1\,.$$

Therefore, the damping coefficient has to satisfy the condition

$$\underline{\underline{d < \frac{4}{9}\sqrt{mk}\,.}}$$

The general solution of the equation of motion is given by (see (5.41))

$$\varphi(t) = C\,\mathrm{e}^{-\xi t}\cos\left(\omega_d\,t - \alpha\right),$$

where

$$\omega_d = \omega\sqrt{1 - \zeta^2} = \frac{1}{2}\sqrt{\frac{k}{m}}\sqrt{1 - \frac{81\,d^2}{16\,mk}}\,.$$

The constants of integration can be calculated from the initial conditions $\varphi(0) = 0$, $\dot{\varphi}(0) = \dot{\varphi}_0$. We obtain $\alpha = \pi/2$ and

$C = \dot{\varphi}_0/\omega_d$. Thus, the solution of the equation of motion is

$$\underline{\underline{\varphi(t) = \frac{\dot{\varphi}_0}{\omega_d} \, e^{-\xi t} \cos\left(\omega_d \, t - \frac{\pi}{2}\right) = \frac{\dot{\varphi}_0}{\omega_d} \, e^{-\xi t} \sin \omega_d \, t.}}$$

E5.6

Example 5.6 Consider again the system shown in Fig. 5.16a. The initial conditions $x(0) = x_0$, $\dot{x}(0) = 0$ and the damping ratio $\zeta = 0.01$ are given.

Determine the energy which is dissipated during the first full cycle.

Solution The initial velocity is zero. Therefore, the total energy E_0 of the system at the beginning of the first cycle is equal to the potential energy V_0 stored in the spring:

$$E_0 = V_0 = \frac{1}{2} k x_0^2.$$

Similarly, the total energy E_1 after the first cycle is given by

$$E_1 = V_1 = \frac{1}{2} k x_1^2,$$

where x_1 is the displacement at time $T_d = 2\pi/\omega_d$.

According to (5.42), (5.36) and (5.40) we have

$$\frac{x_0}{x_1} = e^{\xi T_d} \quad \rightarrow \quad x_1 = x_0 \, e^{-\xi T_d} = x_0 \, e^{-\frac{2\pi\zeta}{\sqrt{1-\zeta^2}}}.$$

This yields the dissipated energy

$$\Delta E = E_0 - E_1 = \frac{1}{2} k x_0^2 - \frac{1}{2} k x_1^2 = \left(1 - e^{-\frac{4\pi\zeta}{\sqrt{1-\zeta^2}}}\right) \frac{1}{2} k x_0^2.$$

With $\zeta = 0.01$ we obtain $\Delta E = 0.13 \cdot \frac{1}{2} k x_0^2$. Thus, 13% of the energy is dissipated during the first full cycle.

5.3

5.3 Forced Vibrations

5.3.1 Undamped Forced Vibrations

We will now investigate the behaviour of a system with one degree of freedom when it is subject to an external force. Let us, as an

illustrative example, consider the system in Fig. 5.20a. The block
is subjected to a harmonic force $F = F_0 \cos \Omega t$, where Ω is the
forcing frequency, i.e., the frequency of the excitation.

We measure the coordinate x from the equilibrium position of
the block in the absence of the force $(F = 0)$. Then we obtain the
equation of motion (see Fig. 5.20b)

$$\downarrow: \quad m\ddot{x} = -kx + F_0 \cos \Omega t \quad \rightarrow \quad m\ddot{x} + kx = F_0 \cos \Omega t. \quad (5.44)$$

In contrast to (5.10), the right-hand side of this equation is *not*
zero: the differential equation is *inhomogeneous*. We introduce the
abbreviations

$$\omega^2 = \frac{k}{m}, \quad x_0 = \frac{F_0}{k} \quad (5.45)$$

where ω is the natural frequency of the system and x_0 is the static
elongation of the spring which is caused by a *constant* force F_0.

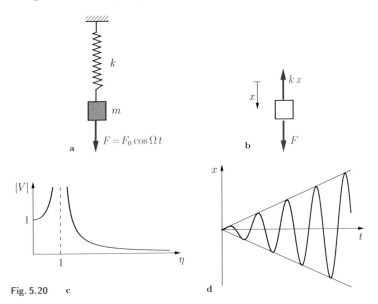

Fig. 5.20   c   d

Then (5.44) becomes

$$\ddot{x} + \omega^2 x = \omega^2 x_0 \cos \Omega t. \quad (5.46)$$

The general solution of this inhomogeneous differential equation is composed of the solution x_h of the homogeneous differential equation ($\ddot{x} + \omega^2 x = 0$) and a particular solution x_p of the inhomogeneous differential equation:

$$x = x_h + x_p .$$

The solution x_h of the homogeneous equation is given by

$$x_h = C \cos(\omega t - \alpha) , \tag{5.47a}$$

see Section 5.2.1. We assume the particular solution x_p to be of the form of the right-hand side:

$$x_p = x_0 V \cos \Omega t . \tag{5.47b}$$

Here, V is a dimensionless quantity which can be determined by inserting x_p into (5.46):

$$- x_0 V \Omega^2 \cos \Omega t + \omega^2 x_0 V \cos \Omega t = \omega^2 x_0 \cos \Omega t$$

$$\rightarrow \quad V = \frac{\omega^2}{\omega^2 - \Omega^2} .$$

We now introduce the non-dimensional *frequency ratio*

$$\eta = \frac{\Omega}{\omega} . \tag{5.48}$$

Then, the quantity V can be written as

$$V = \frac{1}{1 - \eta^2} . \tag{5.49}$$

Thus, the general solution of the differential equation (5.46) is given by

$$x(t) = x_h + x_p = C \cos(\omega t - \alpha) + x_0 V \cos \Omega t . \tag{5.50}$$

The constants of integration C and α can be calculated from the initial conditions. Note that the solution x_h of the homogeneous equation for a real system decays with time due to the ever-present damping (see Section 5.2.3). For this reason the free vibrations (those from the homogeneous solution) are referred to as *transient vibrations*. After a sufficiently long time only the solution x_p

remains. Then the displacement $x(t)$ is represented by

$$x\,(t) = x_p = x_0 V \cos \Omega t\,.$$

These vibrations are called the *steady state vibrations*. The quantity V is the ratio of the amplitude of the vibration to the static elongation x_0. Therefore, V is called the *magnification factor*.

Fig. 5.20c shows the absolute value of the magnification factor V as a function of the frequency ratio η. If the forcing frequency Ω approaches the natural frequency ω of the system $(\eta \to 1)$, the amplitudes of the vibration approach infinity $(V \to \infty)$. This behaviour is called *resonance*. In the case of $\eta \to 0$ we get $V \to 1$ (static displacement in the case of a very small forcing frequency), for $\eta \to \infty$ we have $|V| \to 0$ (vanishing displacement for a very large forcing frequency).

The particular solution (5.47b) is not valid in the case of resonance $(\Omega = \omega)$, since our assumed form does not satisfy the governing equation. In this case,

$$x_p = x_0 \bar{V}\, t \sin \Omega t = x_0 \bar{V}\, t \sin \omega t$$

satisfies the differential equation. We insert the derivatives

$$\dot{x}_p = x_0 \bar{V} \sin \omega t + x_0 \bar{V} \omega t \cos \omega t,$$

$$\ddot{x}_p = 2\, x_0 \bar{V} \omega \cos \omega t - x_0 \bar{V} \omega^2 \, t \sin \omega t$$

into (5.46) to obtain

$$2\, x_0 \bar{V} \omega \cos \omega t - x_0 \bar{V} \omega^2 \, t \sin \omega t + \omega^2 x_0 \bar{V}\, t \sin \omega t = \omega^2 x_0 \cos \omega t$$

$$\to \quad \bar{V} = \frac{\omega}{2}\,.$$

Thus, the particular solution

$$x_p = \frac{1}{2}\, x_0 \, \omega t \sin \omega t$$

represents a "vibration" with a linearly increasing amplitude (Fig. 5.20d).

E5.7

Example 5.7 A block (mass m) is connected to a wall by a spring (spring constant k_1). A second spring (spring constant k_2) connects the block with a circular disk. The disk (radius r, eccentricity e) rotates and causes the block to vibrate. The end point B of the second spring remains in contact with the smooth surface of the disk at all times (Fig. 5.21a).

Determine the circular frequency Ω of the disk so that the amplitude of the steady state vibration of the block is equal to $3\,e$.

Solution We measure the coordinate x from the static equilibrium position of the block when the disk is in the position depicted in Fig. 5.21a. The position of point B is described by the additional coordinate x_B (Fig. 5.21b). The elongation of the second spring is given by $x - x_B$.

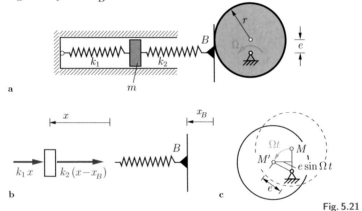

Fig. 5.21

Thus, we obtain the equation of motion

$$\leftarrow: \quad m\ddot{x} = -k_1 x - k_2(x - x_B) \quad \rightarrow \quad m\ddot{x} + (k_1 + k_2)x = k_2 x_B \,.\text{(a)}$$

When the disk rotates, its center moves from the initial position M to the new position M' during the time t (Fig. 5.21c). The displacement of point B coincides with the horizontal component of the displacement of M, i.e., $x_B = e \sin \Omega t$. Substituting into (a) yields

$$\ddot{x} + \omega^2 x = \frac{k_2}{m} e \sin \Omega t \tag{b}$$

with

$$\omega^2 = \frac{k_1 + k_2}{m}. \tag{c}$$

The general solution of this inhomogeneous differential equation is composed of the solution x_h of the homogeneous equation and a particular solution x_p of the inhomogeneous equation. We consider only the steady state vibration and therefore only the solution x_p. We assume x_p to be of the form of the right-hand side of Equation (b):

$$x_p = X \sin \Omega t,$$

where the amplitude X is as yet unknown. Introduction into (b) yields

$$-\Omega^2 X + \omega^2 X = \frac{k_2}{m} e \quad \rightarrow \quad X = \frac{k_2 e}{m(\omega^2 - \Omega^2)}.$$

The graph X versus η was qualitatively given in Fig. 5.20c. The condition $|X| = 3e$ and Equation (c) lead to two frequencies which satisfy the given condition (one frequency for $\eta < 1$ and one frequency for $\eta > 1$):

$$\frac{k_2 e}{m(\omega^2 - \Omega^2)} = \pm 3\,e \quad \rightarrow \quad \begin{cases} \underline{\underline{\Omega_1^2}} = \omega^2 - \dfrac{k_2}{3\,m} = \underline{\dfrac{3\,k_1 + 2\,k_2}{3\,m}}, \\[3mm] \underline{\underline{\Omega_2^2}} = \omega^2 + \dfrac{k_2}{3\,m} = \underline{\dfrac{3\,k_1 + 4\,k_2}{3\,m}}. \end{cases}$$

5.3.2 Damped Forced Vibrations

In this section we will investigate damped forced vibrations, restricting ourselves to viscous damping. Thereby we will consider the following three cases.

Case 1: *Excitation through a force* or *via a spring*

Let us first consider a system which consists of a block, a spring and a damper as shown in Fig. 5.22a. The block is subjected to a harmonic force $F = F_0 \cos \Omega t$ where Ω is a given constant forcing frequency. The equation of motion is obtained as

$$\uparrow:\quad m\ddot{x} = -kx - d\dot{x} + F_0 \cos \Omega t \quad \rightarrow \quad m\ddot{x} + d\dot{x} + kx = F_0 \cos \Omega t\,.$$
$$(5.51)$$

If we introduce the abbreviations

$$2\xi = \frac{d}{m}, \quad \omega^2 = \frac{k}{m}, \quad x_0 = \frac{F_0}{k} \qquad (5.52)$$

(cf. (5.31) and (5.45)) we obtain the differential equation

$$\ddot{x} + 2\,\xi\,\dot{x} + \omega^2\,x = \omega^2\,x_0 \cos \Omega t\,. \qquad (5.53)$$

We will now consider the system in Fig. 5.22b where the free end of the spring is forced to move according to $x_S = x_0 \cos \Omega t$. Note that no external force acts on the block. Then the elongation of the spring is given by $x_S - x$ which leads to the equation of motion for the block:

$$\uparrow:\quad m\ddot{x} = k\,(x_S - x) - d\,\dot{x} \quad \rightarrow \quad m\ddot{x} + d\,\dot{x} + k\,x = k\,x_0 \cos \Omega t\,.$$

With the abbreviations (5.52) we again obtain (5.53):

$$\ddot{x} + 2\,\xi\,\dot{x} + \omega^2\,x = \omega^2\,x_0 \cos \Omega t\,.$$

Thus, the motions of the blocks of both systems are described by the same equation.

Case 2: *Excitation via a damper*

Fig. 5.22c shows a system where the free end of the damper undergoes the prescribed motion $x_\zeta = x_0 \sin \Omega t$. The damping force in the dashpot is proportional to the relative velocity $\dot{x}_\zeta - \dot{x}$ between the piston and the encasement. Thus, the equation of motion is given by

$$\uparrow:\quad m\ddot{x} = -k\,x + d\,(\dot{x}_\zeta - \dot{x}) \quad \rightarrow \quad m\ddot{x} + d\dot{x} + k\,x = d\,\Omega x_0 \cos \Omega t\,.$$

With the abbreviations

$$2\,\xi = \frac{d}{m}, \quad \omega^2 = \frac{k}{m}, \quad \zeta = \frac{\xi}{\omega}, \quad \eta = \frac{\Omega}{\omega} \qquad (5.54)$$

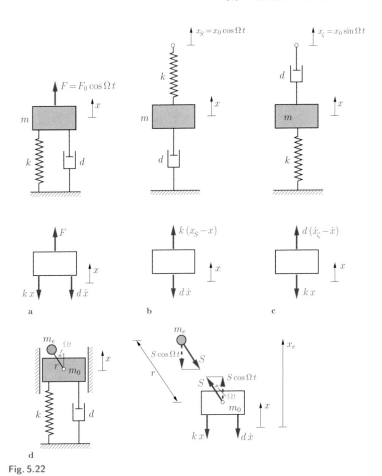

Fig. 5.22

(cf. (5.31), (5.36) and (5.48)) we obtain

$$\ddot{x} + 2\,\xi\,\dot{x} + \omega^2\,x = 2\,\xi\Omega\,x_0 \cos\Omega t$$

$$\rightarrow \quad \ddot{x} + 2\,\xi\,\dot{x} + \omega^2\,x = 2\zeta\eta\omega^2 x_0 \cos\Omega t\,. \tag{5.55}$$

Case 3: *Unbalanced Rotation*

The block (mass m_0) in Fig. 5.22d is forced to vibrate by a rotating eccentric mass m_e. The constant frequency Ω of the rotation is given. The positions of the block and of the eccentric mass, respectively, are described by the coordinates x and x_e, counted upwards from the same position. Then the relation

$$x_e = x + r \cos \Omega t \quad \rightarrow \quad \ddot{x}_e = \ddot{x} - r \, \Omega^2 \cos \Omega t$$

can be found from the figure. We denote the force in the bar connecting the block and the eccentric mass by S. The equations of motion in the vertical direction for the masses m_e and m_0 are then found to be (see the free-body diagrams)

$$\uparrow: \quad m_e \ddot{x}_e = - S \cos \Omega t,$$

$$\uparrow: \quad m_0 \ddot{x} \quad = - k \, x - d \, \dot{x} + S \cos \Omega t \,.$$

We eliminate S and insert $\ddot{x}_e$ from above to obtain

$$(m_0 + m_e)\ddot{x} + d \, \dot{x} + k \, x = m_e r \, \Omega^2 \cos \Omega t \,.$$

Now we introduce the abbreviations

$$m = m_0 + m_e, \quad x_0 = \frac{m_e}{m} r \tag{5.56}$$

and use (5.54). This leads to the equation of motion for m_0:

$$\ddot{x} + 2 \, \xi \, \dot{x} + \omega^2 \, x = \omega^2 \, \eta^2 x_0 \cos \Omega t \,. \tag{5.57}$$

The three equations of motion (5.53, 5.55 and 5.57) differ only in the factor in front of the cosine function on the right-hand sides. Therefore, with $\zeta = \xi/\omega$ they can be written as a single equation:

$$\ddot{x} + 2 \, \zeta \omega \, \dot{x} + \omega^2 \, x = x_0 E \omega^2 \, \cos \Omega t \,. \tag{5.58a}$$

Here, the value of E has to be chosen according to the type of forcing:

Case 1: $E = 1$,

Case 2: $E = 2 \, \zeta \, \eta,$ \qquad\qquad\qquad (5.58b)

Case 3: $E = \eta^2 \,.$

The general solution of (5.58a) is composed of the solution x_h of the homogeneous differential equation and a particular solution x_p of the inhomogeneous differential equation (cf. the undamped forced vibrations). According to Section 5.3.1 the solution x_h decays exponentially with time. Therefore this solution which represents the transient vibrations may be neglected compared with the particular solution x_p after a sufficiently long time.

We assume the steady state vibrations x_p to be of the form of the right-hand side of the differential equation (cf. the undamped vibrations). However, we have to also allow for a *phase angle* φ between the applied force and the response:

$$x_p = x_0\, V \cos(\Omega t - \varphi)\,. \tag{5.59}$$

If we insert

$$x_p = x_0\, V\,(\cos \Omega t \cos \varphi + \sin \Omega t \sin \varphi),$$

$$\dot{x}_p = x_0\, V\Omega\,(-\sin \Omega t \cos \varphi + \cos \Omega t \sin \varphi),$$

$$\ddot{x}_p = x_0\, V\Omega^2\,(-\cos \Omega t \cos \varphi - \sin \Omega t \sin \varphi)$$

into the differential equation (5.58a) we obtain

$$x_0 V\Omega^2\,(-\cos \Omega t \cos \varphi - \sin \Omega t \sin \varphi)$$

$$+\,2\,\zeta\,x_0 V\Omega\omega\,(-\sin \Omega t \cos \varphi + \cos \Omega t \sin \varphi)$$

$$+\,x_0\,V\omega^2\quad(\cos \Omega t \cos \varphi + \sin \Omega t \sin \varphi) = x_0 E\omega^2 \cos \Omega t\,.$$

With the frequency ratio $\eta = \Omega/\omega$ this leads to

$$(-V\eta^2 \cos \varphi + 2\,\zeta V\eta \sin \varphi + V \cos \varphi - E)\cos \Omega t$$

$$+\,(-V\eta^2 \sin \varphi - 2\,\zeta V\eta \cos \varphi + V \sin \varphi)\sin \Omega t = 0\,.$$

This equation has to be satisfied at all times t. Therefore the terms in both parentheses have to vanish:

$$V\,(-\eta^2 \cos \varphi + 2\,\zeta\,\eta \sin \varphi + \cos \varphi) = E, \tag{5.60a}$$

$$-\eta^2 \sin \varphi - 2\,\zeta\,\eta \cos \varphi + \sin \varphi = 0\,. \tag{5.60b}$$

We can calculate the *phase angle* φ from (5.60b):

$$\tan \varphi = \frac{2 \zeta \eta}{1 - \eta^2} .$$
(5.61)

Using the standard trigonometric relations

$$\sin \varphi = \frac{\tan \varphi}{\sqrt{1 + \tan^2 \varphi}}, \quad \cos \varphi = \frac{1}{\sqrt{1 + \tan^2 \varphi}}$$

we obtain the *magnification ratio* V, also called *frequency response* from (5.60a):

$$V = \frac{E}{\sqrt{(1 - \eta^2)^2 + 4 \zeta^2 \eta^2}} .$$
(5.62)

Corresponding to the three values of E in (5.58b) we obtain three different magnification factors V_i. They are displayed in the Figs. 5.23a–c for several values of the damping factor ζ. If the excitation of the system is due to a force acting on the block or if the system is forced into vibration via the spring (case 1: $E = 1$), the magnification factor is given by V_1 (Fig. 5.23a) where

$$V_1 (0) = 1, \quad V_1 (1) = \frac{1}{2 \zeta}, \quad V_1 (\eta \to \infty) \to 0 .$$

If $\zeta^2 \leq 0.5$ the curves take on their maximum values $V_{1m} = 1/(2 \zeta \sqrt{1 - \zeta^2})$ at $\eta_m = \sqrt{1 - 2 \zeta^2}$. Note that the maximum values are *not* located at the positions of the natural frequencies of the damped vibrations. In the case of small damping ($\zeta \ll 1$) we get $\eta_m \approx 1$ and $V_{1m} \approx 1/2\zeta$ (resonance); in the limit $\zeta \to 0$ the magnification factor V_1 converges towards the magnification factor (5.49). If $\zeta^2 > 0.5$ the curves decrease monotonously towards zero.

In the case of the forcing via the damper (case 2: $E = 2 \zeta \eta$) we obtain V_2 (Fig. 5.23b). Here,

$$V_2 (0) = 0, \quad V_2 (1) = 1, \quad V_2 (\eta \to \infty) \to 0 .$$

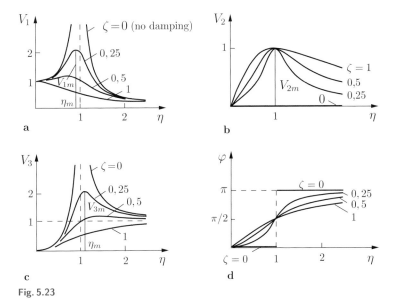

Fig. 5.23

The maximum value $V_{2m} = 1$ is independent of ζ and is always located at $\eta_m = 1$.

An unbalanced rotation (case 3: $E = \eta^2$) is represented by V_3 (Fig. 5.23c) where

$$V_3\left(0\right) = 0, \quad V_3\left(1\right) = \frac{1}{2\,\zeta}, \quad V_3\left(\eta \to \infty\right) \to 1\,.$$

If $\zeta^2 \le 0.5$ the curves attain their maximum values $V_{3m} = 1/(2\,\zeta\sqrt{1 - \zeta^2})$ at $\eta_m = 1/\sqrt{1 - 2\zeta^2}$. For $\zeta^2 > 0.5$ they increase monotonically towards 1. Small damping leads to $\eta_m \approx 1$, $V_{3m} \approx 1/2\zeta$ (cf. case 1).

The phase angle φ is independent of E, see (5.61). Therefore it is the same for the three cases. It represents the delay of the response relative to the excitation. Fig. 5.23d shows φ as a function of the frequency ratio η. Here we have

$$\varphi\left(0\right) = 0, \quad \varphi\left(1\right) = \pi/2, \quad \varphi\left(\eta \to \infty\right) \to \pi\,.$$

In the case of a small forcing frequency ($\eta \ll 1$) excitation and response are "in phase" ($\varphi \approx 0$); in the case of a large forcing frequency ($\eta \gg 1$) they are 180° out of phase ($\varphi \approx \pi$). In the limit $\zeta \to 0$ we have a discontinuity at $\eta = 1$, namely a jump of the phase angle φ from 0 to π.

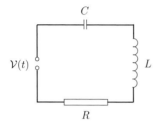

Fig. 5.24

Table 5.1

Mechanical system		Elektrical circuit	
x	displacement	Q	electric charge
$v = \dot{x}$	velocity	$I = \dot{Q}$	electric current
m	mass	L	inductance
d	damping coefficient	R	resistance
k	spring constant	$1/C$	1/capacitance
F	force	$\mathcal{V}$	voltage

Finally we want to show that there exists a relationship between an electrical circuit and a vibrating mechanical system. As an example let us consider the electrical circuit shown in Fig. 5.24. It consists of a capacitor (capacitance C), an inductor (inductance L) and a resistor (resistance R). If a voltage $\mathcal{V}(t) = \mathcal{V}_0 \cos \Omega t$ is applied, the change of the electric charge Q (electric current $I = \dot{Q}$) is described by

$$L\ddot{Q} + R\dot{Q} + \frac{1}{C} Q = \mathcal{V}_0 \cos \Omega t \,.$$

If we replace L by m, R by $d, 1/C$ by $k, \mathcal{V}_0$ by F_0 and Q by x, we obtain the equation of motion (5.51) of a vibrating mechanical system.

Thus, there exists an analogy between an electrical circuit and a mechanical system. The corresponding quantities are shown in Table 5.1.

Example 5.8 A device to measure vibrations is schematically shown in Fig. 5.25a. The encasement is forced into harmonic vibration according to $x_G = x_0 \cos \Omega t$. The amplitudes of the forcing, x_0, and of the response are supposed to coincide for a large range of the forcing frequency Ω and for arbitrary values of the damping coefficient.

Determine the required parameters k and m.

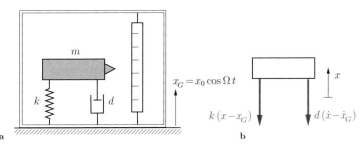

Fig. 5.25

Solution We introduce the coordinate x, measured from a fixed point as shown in Fig. 5.25b. Then the displacement and the velocity, respectively, of the mass relative to the encasement are given by $x - x_G$ and $\dot{x} - \dot{x}_G$. This yields the equation of motion

$$\uparrow: \quad m\ddot{x} = -k(x - x_G) - d(\dot{x} - \dot{x}_G) \,. \tag{a}$$

The device measures the displacement $x_r = x - x_G$ relative to the encasement. Using $\dot{x}_r = \dot{x} - \dot{x}_G$, $\ddot{x}_r = \ddot{x} - \ddot{x}_G$ and $\ddot{x}_G = -x_0\Omega^2 \cos \Omega t$, we obtain from (a)

$$m\ddot{x}_r + d\,\dot{x}_r + k\,x_r = m\Omega^2\,x_0\cos\Omega t\,.$$

Dividing by m and using the abbreviations (5.54) leads to a differential equation which is analogous to (5.57), case 3:

$$\ddot{x}_r + 2\xi\,\dot{x}_r + \omega^2 x_r = \omega^2\,\eta^2\,x_0\cos\Omega t\,.$$

The steady state solution is given by (5.59):

$$x_r = x_p = x_0\,V_3\cos\left(\Omega t - \varphi\right).$$

The measured amplitude and the amplitude of the forcing coincide if the magnification factor is equal to 1, i.e. $V_3 = 1$. Fig. 5.23c shows that this requirement is approximately satisfied for $\eta \gg 1$, independent of the damping ratio ζ. Thus,

$$\omega^2 \ll \Omega^2 \quad \rightarrow \quad \underline{\frac{k}{m} \ll \Omega^2}\,.$$

Hence, the natural frequency of the undamped vibrations has to be much smaller than the forcing frequency (soft spring!).

5.4 Systems with two Degrees of Freedom

5.4.1 Free Vibrations

In the following we will investigate free vibrations of systems with two degrees of freedom. As an example let us consider the system shown in Fig. 5.26a which consists of two blocks and two springs. We introduce the two coordinates x_1 and x_2 which describe the positions of the blocks. The coordinates are measured from the equilibrium positions of the blocks (Fig. 5.26b).

To obtain the equations of motion we make use of Lagrange's equations of the second kind. The kinetic and the potential energy, respectively, are given by

$$T = \tfrac{1}{2}m_1\,\dot{x}_1^2 + \tfrac{1}{2}m_2\,\dot{x}_2^2,$$
$$V = \tfrac{1}{2}k_1\,x_1^2 + \tfrac{1}{2}k_2\,(x_2 - x_1)^2\,. \tag{5.63}$$

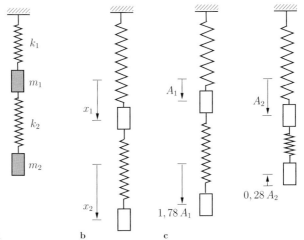

Fig. 5.26 a   b   c

With the Lagrangian $L = T - V$ we obtain the equations of motion from (4.36):

$$m_1 \ddot{x}_1 + k_1 x_1 - k_2(x_2 - x_1) = 0,$$

$$m_2 \ddot{x}_2 + k_2 (x_2 - x_1) = 0$$

or

$$m_1 \ddot{x}_1 + (k_1 + k_2)x_1 - k_2 x_2 = 0,$$

$$m_2 \ddot{x}_2 - k_2 x_1 + k_2 x_2 = 0 \,. \tag{5.64}$$

We assume the solution of this system of two coupled homogeneous differential equations of second order with constant coefficients to be of the form

$$x_1 = A \cos \omega t, \quad x_2 = C \cos \omega t \,, \tag{5.65}$$

where A, C and ω are as yet unknown. We insert (5.65) into (5.64) and obtain the homogeneous algebraic system of equations

$$(k_1 + k_2 - m_1 \omega^2)A - k_2 C = 0,$$

$$-k_2 A + (k_2 - m_2 \omega^2) C = 0 \tag{5.66}$$

for the constants A and C. The trivial solution $A = C = 0$ leads to $x_1 = x_2 = 0$ (see (5.65)). In order to find a nontrivial solution, the determinant of the matrix of the coefficients has to vanish:

$$\Delta(\omega) = \begin{vmatrix} k_1 + k_2 - m_1\,\omega^2 & -k_2 \\ -k_2 & k_2 - m_2\,\omega^2 \end{vmatrix} = 0\,. \tag{5.67}$$

This yields the *characteristic equation*

$$(k_1 + k_2 - m_1\,\omega^2)(k_2 - m_2\,\omega^2) - k_2^2 = 0 \tag{5.68}$$

or

$$m_1\,m_2\,\omega^4 - (m_1\,k_2 + m_2\,k_1 + m_2\,k_2)\,\omega^2 + k_1\,k_2 = 0\,. \tag{5.69}$$

The characteristic equation is a quadratic equation for ω^2. Its solutions ω_1^2 and ω_2^2 are positive according to Decartes' rule of signs:

$$\omega_1^2\,\omega_2^2 = \frac{k_1\,k_2}{m_1\,m_2} > 0, \quad \omega_1^2 + \omega_2^2 = \frac{m_1\,k_2 + m_2\,k_1 + m_2\,k_2}{m_1\,m_2} > 0\,. \tag{5.70}$$

The two values ω_1 and ω_2 are the *two* natural frequencies (eigen-frequencies) of the system. We will number them so that $\omega_2 > \omega_1$.

The constants A and C are not independent. To find the relationship between these constants we insert one of the natural frequencies, for example ω_1, into the first equation of (5.66). Then we find the ratio between the corresponding amplitudes A_1 and C_1:

$$(k_1 + k_2 - m_1\,\omega_1^2)A_1 - k_2\,C_1 = 0$$

$$\rightarrow \quad \mu_1 = \frac{C_1}{A_1} = \frac{k_1 + k_2 - m_1\,\omega_1^2}{k_2}\,. \tag{5.71}$$

Note that the same result is obtained if we insert ω_1 into the second equation of (5.66). With (5.71). Equation (5.65) becomes

$$x_1 = A_1\cos\omega_1 t, \quad x_2 = \mu_1 A_1\cos\omega_1 t\,. \tag{5.72}$$

If we insert the second natural frequency ω_2 into one of the equations (5.66) we obtain

$$\mu_2 = \frac{C_2}{A_2} = \frac{k_1 + k_2 - m_1\,\omega_2^2}{k_2} \tag{5.73}$$

and

$$x_1 = A_2 \cos\omega_2 t, \quad x_2 = \mu_2 A_2 \cos\omega_2 t\,. \tag{5.74}$$

We find two additional independent solutions of (5.64) if we replace the cosine function in (5.72) and (5.74), respectively, by the sine function. The general solution of (5.64) is a linear combination of these four independent solutions. Thus, it is given by

$$x_1 = A_1 \cos\omega_1 t + B_1 \sin\omega_1 t + A_2 \cos\omega_2 t + B_2 \sin\omega_2 t,$$
$$\tag{5.75}$$
$$x_2 = \mu_1\,A_1 \cos\omega_1 t + \mu_1\,B_1 \sin\omega_1 t + \mu_2\,A_2 \cos\omega_2 t + \mu_2\,B_2 \sin\omega_2 t\,.$$

The four constants of integration can be calculated from the initial conditions. Note that ω_1, ω_2, μ_1 and μ_2 are independent of the initial conditions.

If the initial conditions are chosen appropriately all the integration constants except one in the general solution (5.75) are zero. Then both blocks move in the form of a cosine (or sine) function only with the first natural frequency or only with the second one (cf. (5.72) or (5.74)). The blocks attain their maximum displacements simultaneously and pass their equilibrium positions simultaneously. Such motions are called *principal modes of vibration* or *eigenmodes*.

We will now continue the example using the parameters $m_1 = m$, $m_2 = 2\,m$, and $k_1 = k_2 = k$. Then (5.69) yields the characteristic equation

$$2\,m^2\omega^4 - 5c\,m\omega^2 + k^2 = 0 \tag{5.76}$$

with the solutions

$$\omega_1^2 = \frac{1}{4}\,(5 - \sqrt{17})\,\frac{k}{m} = 0.219\,\frac{k}{m},$$
$$\tag{5.77}$$
$$\omega_2^2 = \frac{1}{4}\,(5 + \sqrt{17})\,\frac{k}{m} = 2.28\,\frac{k}{m}\,.$$

Hence, the natural frequencies are

$$\omega_1 = 0.468\sqrt{\frac{k}{m}}, \quad \omega_2 = 1.51\sqrt{\frac{k}{m}} \tag{5.78}$$

and the ratios between the amplitudes follow from (5.71) and (5.73):

$$\begin{aligned}
\mu_1 &= \frac{2\,k - m\omega_1^2}{k} = 2 - \frac{m}{k}\,\omega_1^2 = 1.78, \\
\mu_2 &= \frac{2\,k - m\omega_2^2}{k} = 2 - \frac{m}{k}\,\omega_2^2 = -0.28.
\end{aligned} \tag{5.79}$$

Let us first assume that both blocks vibrate with the first natural frequency ω_1 (first principal mode of vibration). Then both displacements x_1 and x_2 always have the same algebraic sign since $\mu_1 > 0$: both blocks always move in the same direction; they vibrate "in phase". On the other hand, if the vibration takes place with the second natural frequency ω_2 (second principal mode of vibration) the algebraic signs of x_1 and x_2 are always different because $\mu_2 < 0$: both blocks always move in opposite directions; they vibrate 180° out of phase. Fig. 5.26c illustrates the displacements at a fixed time, i.e. the mode shapes, for both cases. A vibration with arbitrary initial conditions is obtained through a superposition of the two principal modes.

The equations of motion (5.64) are coupled in the coordinates x_1 and x_2. If we introduce the matrices

$$\boldsymbol{m} = \begin{bmatrix} m_1 & 0 \\ 0 & m_2 \end{bmatrix}, \quad \boldsymbol{k} = \begin{bmatrix} k_1 + k_2 & -k_2 \\ -k_2 & k_2 \end{bmatrix} \tag{5.80}$$

and the column vectors

$$\boldsymbol{x} = \begin{bmatrix} x_1 \\ x_2 \end{bmatrix}, \quad \ddot{\boldsymbol{x}} = \begin{bmatrix} \ddot{x}_1 \\ \ddot{x}_2 \end{bmatrix} \tag{5.81}$$

we can write them in the form of a matrix equation:

$$\boldsymbol{m}\ddot{\boldsymbol{x}} + \boldsymbol{k}\,\boldsymbol{x} = \boldsymbol{0}. \tag{5.82}$$

Frequently the equations of motion are also coupled in the accelerations $\ddot{x}_1$ and $\ddot{x}_2$ (cf. Example 5.9). Then the matrix $\boldsymbol{m}$ in (5.82) is no longer a diagonal matrix. Thus in the general case the matrices $\boldsymbol{m}$ and $\boldsymbol{k}$ are given by

$$\boldsymbol{m} = \begin{bmatrix} m_{11} & m_{12} \\ m_{21} & m_{22} \end{bmatrix}, \quad \boldsymbol{k} = \begin{bmatrix} k_{11} & k_{12} \\ k_{21} & k_{22} \end{bmatrix}. \tag{5.83}$$

Note that the type of the coupling depends on the choice of the coordinates, *not* on the mechanical system. Coordinates can also be chosen where $\boldsymbol{k}$ is diagonal.

To solve the differential equation (5.82) a solution in the form of $\boldsymbol{x} = \boldsymbol{A}\,\mathrm{e}^{\mathrm{i}\omega t}$ may be assumed. This is equivalent to (5.65); via the characteristic equation it also leads to the natural frequencies and the principal modes.

Example 5.9   A cantilever beam (with negligible mass, flexural rigidity EI) carries two mass points $(m_1 = 2\,m, m_2 = m)$ as shown in Fig. 5.27a.

E5.9

Determine the natural frequencies and the principal modes of vibration.

Solution   The position of the system is uniquely determined by the deflections w_1 and w_2 of the two mass points (Fig. 5.27b). Therefore the system has two degrees of freedom.

If the mass points are displaced they are subjected to the restoring forces F_1 and F_2 (Fig. 5.27c). Thus, the equations of motion are

$$m_1\,\ddot{w}_1 = -\,F_1, \quad m_2\,\ddot{w}_2 = -\,F_2\,. \tag{a}$$

The relationship between the forces F_1, F_2 and the deflections w_1, w_2 can be obtained with the aid of the influence coefficients α_{ik} (see Volume 2, Section 6.4):

$$\begin{aligned} w_1 &= \alpha_{11}\,F_1 + \alpha_{12}\,F_2, \\ w_2 &= \alpha_{21}\,F_1 + \alpha_{22}\,F_2\,. \end{aligned} \tag{b}$$

Here, α_{ik} is the deflection at position i due to a unit force at position k. If we insert the forces according to (a) into (b) we get

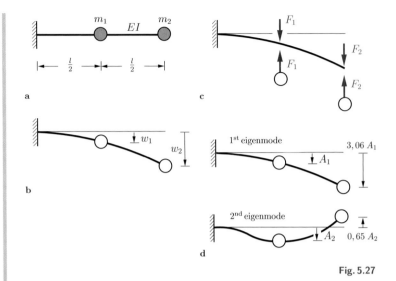

Fig. 5.27

$$\alpha_{1\,1}\,m_1\ddot{w}_1 + \alpha_{1\,2}\,m_2\,\ddot{w}_2 + w_1 = 0,$$

$$\alpha_{2\,1}\,m_1\ddot{w}_1 + \alpha_{2\,2}\,m_2\,\ddot{w}_2 + w_2 = 0\,.$$

With the influence coefficients

$$\alpha_{1\,1} = \frac{l^3}{24\,EI}, \quad \alpha_{2\,2} = \frac{l^3}{3\,EI}, \quad \alpha_{1\,2} = \alpha_{2\,1} = \frac{5\,l^3}{48\,EI},$$

the abbreviation $\alpha = \dfrac{l^3}{48\,EI}$ and the parameters $m_1 = 2\,m$, $m_2 = m$ we obtain

$$4\,\alpha m\,\ddot{w}_1 + \ 5\,\alpha m\,\ddot{w}_2 \ + w_1 = 0,$$

$$10\,\alpha m\,\ddot{w}_1 + 16\,\alpha m\,\ddot{w}_2 + w_2 = 0\,.$$

Now we assume a solution of the form

$$w_1 = A\cos\omega t, \quad w_2 = C\cos\omega t$$

which leads to the system of linear equations

$$(1 - 4\,\alpha m\,\omega^2)A - 5\,\alpha m\,\omega^2\,C = 0,$$

$$-\,10\,\alpha m\,\omega^2 A + (1 - 16\,\alpha m\,\omega^2)C = 0\,.$$

(c)

The characteristic equation

$$14\,\alpha^2\,m^2\,\omega^4 - 20\,\alpha m\,\omega^2 + 1 = 0$$

yields the natural frequencies:

$$\omega_1^2 = \frac{10 - \sqrt{86}}{14\,\alpha m} = 0.0519/(\alpha m) \quad \rightarrow \quad \underline{\underline{\omega_1 = 0.23/\sqrt{\alpha m}}},$$

$$\omega_2^2 = \frac{10 + \sqrt{86}}{14\,\alpha m} = 1.377/(\alpha m) \quad \rightarrow \quad \underline{\underline{\omega_2 = 1.17/\sqrt{\alpha m}}}.$$

We obtain the ratios between the amplitudes by inserting the natural frequencies into (c):

$$\mu_1 = \frac{C_1}{A_1} = \frac{1 - 4\,\alpha m\,\omega_1^2}{5\,\alpha m\,\omega_1^2} = 3.06\,,$$

$$\mu_2 = \frac{C_2}{A_2} = \frac{1 - 4\,\alpha m\,\omega_2^2}{5\,\alpha m\,\omega_2^2} = -0.65\,.$$

The principal modes of the vibration (Fig. 5.27d) are given by these ratios. The mass points vibrate in phase in the first principal mode and 180° out of phase in the second one.

Example 5.10 A frame consists of two rigid horizontal beams (masses m_1 and m_2) which are rigidly attached to two elastic columns (with negligible masses) as shown in Fig. 5.28a. Given: $m_1 = 1000$ kg, $m_2 = \frac{3}{2}m_1$, $E = 210$ GPa, $I = 5100$ cm^4, $h = 4.5$ m.

E5.10

Determine the natural frequencies and the principal modes of vibration.

Solution Since the columns are elastic the beams undergo horizontal displacements (cf. Example 5.4). We denote these displacements by w_1 and w_2 (Fig. 5.28b). The frame is replaced by the equivalent model in Fig. 5.28c where the spring constants

$$k_1 = k_2 = k = \frac{24\,EI}{h^3} \tag{a}$$

are known from Example 5.4 (two columns in parallel). Thus, the kinetic energy T and the potential energy V are given by

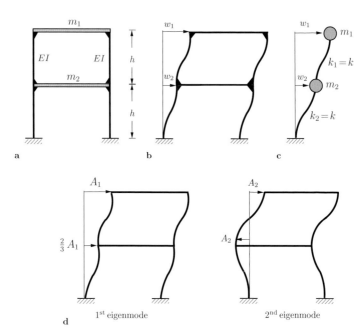

Fig. 5.28

(cf. (5.63))

$$T = \frac{1}{2}\,m_1\,\dot{w}_1^2 + \frac{1}{2}\,m_2\,\dot{w}_2^2 = \frac{1}{2}\,m(\dot{w}_1^2 + \frac{3}{2}\,\dot{w}_2^2),$$

$$V = \frac{1}{2}\,k\,(w_1 - w_2)^2 + \frac{1}{2}\,kw_2^2,$$

where $m = m_1$. With $L = T - V$, Lagrange's equations (4.36) yield

$$m\,\ddot{w}_1 + k\,w_1 - k\,w_2 = 0,$$
$$3\,m\,\ddot{w}_2 - 2\,k\,w_1 + 4\,k\,w_2 = 0\,.$$

Assuming a solution in the form

$$w_1 = A\,e^{i\omega t}, \quad w_2 = C\,e^{i\omega t}$$

we obtain the system of linear equations

$$(k - m\,\omega^2)A - k\,C = 0,$$

$$- 2\,k\,A + (4\,k - 3\,m\,\omega^2)\,C = 0\,.$$

(b)

The characteristic equation

$$3\,m^2\,\omega^4 - 7\,k\,m\,\omega^2 + 2\,k^2 = 0$$

has the solutions

$$\omega_1^2 = \frac{k}{3\,m}, \quad \omega_2^2 = \frac{2\,k}{m}\,.$$

Using (a) we finally get

$$\omega_1 = 2\sqrt{\frac{2\,EI}{m\,h^3}} = 30.7 \text{ s}^{-1}, \quad \omega_2 = 4\sqrt{\frac{3\,EI}{m\,h^3}} = 75.1 \text{ s}^{-1}\,.$$

These circular frequencies are equivalent to the frequencies $f_1 = 4.9$ Hz and $f_2 = 12.0$ Hz.

The ratios between the amplitudes follow after substituting the natural frequencies into (b):

$$\mu_1 = \frac{C_1}{A_1} = 1 - \frac{m_1}{k}\,\omega_1^2 = \frac{2}{3}\,,$$

$$\mu_2 = \frac{C_2}{A_2} = 1 - \frac{m_1}{k}\,\omega_2^2 = -1\,.$$

The mode shapes are shown in Fig. 5.28d.

5.4.2 Forced Vibrations

We will investigate forced vibrations of a two degrees of freedom system with the aid of an example only. Let us consider the system in Fig. 5.29a. A block (mass m_1) is supported by two springs (spring constant of each spring $k_1/2$). It can move in the vertical direction and is subjected to a harmonic force $F = F_0 \cos \Omega t$. A second block (mass m_2) is suspended from the first one by another spring (spring constant k_2). We measure the coordinates x_1 and x_2 from the equilibrium positions ($F = 0$) of m_1 and m_2. Then

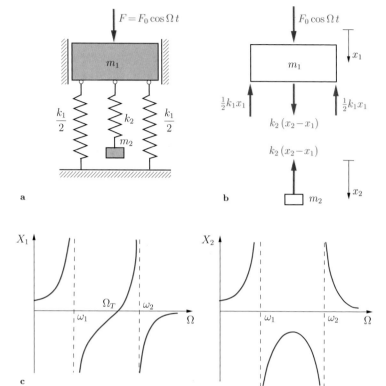

Fig. 5.29

we obtain the equations of motion (see Fig. 5.29b)

$$m_1 \ddot{x}_1 = -2 \cdot \tfrac{1}{2} k_1 x_1 + k_2 (x_2 - x_1) + F_0 \cos \Omega t,$$

$$m_2 \ddot{x}_2 = -k_2 (x_2 - x_1)$$

or

$$m_1 \ddot{x}_1 + (k_1 + k_2)x_1 - k_2 x_2 = F_0 \cos \Omega t,$$

$$m_2 \ddot{x}_2 - k_2 x_1 + k_2 x_2 = 0.$$

(5.84)

This is a system of *inhomogeneous* differential equations of second order. The general solution x_j $(j = 1, 2)$ is composed of the solution x_{jh} of the homogeneous differential equations and a particular solution x_{jp} of the inhomogeneous differential equations: $x_j = x_{jh} + x_{jp}$. Since the solution x_{jh} decays due to the damping which is always present in real systems (cf. Section 5.3.2) we will consider only the particular solution x_{jp}. We assume this solution of (5.84) to be of the form of the right-hand side:

$$x_{1p} = X_1 \cos \Omega t, \quad x_{2p} = X_2 \cos \Omega t.$$

Introduction into (5.84) leads to

$$[(k_1 + k_2 - m_1 \Omega^2)X_1 - k_2 X_2] \cos \Omega t = F_0 \cos \Omega t,$$

$$[- k_2 X_1 + (k_2 - m_2 \Omega^2)X_2] \cos \Omega t = 0$$

and thus to the inhomogeneous system of linear equations

$$\left(\frac{k_1 + k_2}{m_1} - \Omega^2 \right) X_1 - \frac{k_2}{m_1} X_2 = \frac{F_0}{m_1},$$

$$- \frac{k_2}{m_2} X_1 + \left(\frac{k_2}{m_2} - \Omega^2 \right) X_2 = 0 \tag{5.85}$$

for the amplitudes X_1 and X_2. Equations (5.85) have the solution

$$X_1 = \frac{\dfrac{F_0}{m_1} \left(\dfrac{k_2}{m_2} - \Omega^2 \right)}{\Delta(\Omega)}, \quad X_2 = \frac{\dfrac{F_0}{m_1} \dfrac{k_2}{m_2}}{\Delta(\Omega)}, \tag{5.86}$$

where

$$\Delta(\Omega) = \left(\frac{k_1 + k_2}{m_1} - \Omega^2 \right) \left(\frac{k_2}{m_2} - \Omega^2 \right) - \frac{k_2^2}{m_1 m_2} \tag{5.87}$$

is the determinant of the matrix of the coefficients of (5.85).

Equation (5.87) can be simplified. According to Section 5.4.1 the natural frequencies ω_1 and ω_2 of the *free* vibrations follow from the characteristic equation

$$\Delta(\omega) = \left(\frac{k_1 + k_2}{m_1} - \omega^2 \right) \left(\frac{k_2}{m_2} - \omega^2 \right) - \frac{k_2^2}{m_1 m_2} = 0. \tag{5.88}$$

It has the solutions ω_1^2 and ω_2^2 and can therefore be written in the form

$$\Delta(\omega) = (\omega^2 - \omega_1^2)(\omega^2 - \omega_2^2) = 0. \tag{5.89}$$

The comparison of (5.87) and (5.88) yields with (5.89) the relation

$$\Delta(\Omega) = (\Omega^2 - \omega_1^2)(\Omega^2 - \omega_2^2).$$

Thus, (5.86) becomes

$$X_1 = \frac{\dfrac{F_0}{m_1}\left(\dfrac{k_2}{m_2} - \Omega^2\right)}{(\Omega^2 - \omega_1^2)(\Omega^2 - \omega_2^2)}, \qquad X_2 = \frac{\dfrac{F_0}{m_1}\dfrac{k_2}{m_2}}{(\Omega^2 - \omega_1^2)(\Omega^2 - \omega_2^2)}. \qquad (5.90)$$

The graphs of the amplitudes X_1 and X_2 versus the forcing frequency Ω are qualitatively shown in Fig. 5.29c. The amplitudes approach infinity (denominator equal to zero, undamped vibrations) for $\Omega = \omega_1$ and $\Omega = \omega_2$: there are two resonant frequencies.

If the forcing frequency is equal to $\Omega_T = \sqrt{k_2/m_2}$, then the amplitude X_1 vanishes: $X_1 = 0$. In this case the mass m_1 does *not* move; only the mass m_2 vibrates. The frequency of this vibration is given by the eigenfrequency $\sqrt{k_2/m_2}$ of the single-degree-of-freedom system consisting of the spring k_2 and the mass m_2. This phenomenon can be exploited if the displacement of the mass m_1 and thus the forces which are transferred to the ground from the springs are to be kept small. The spring-mass system $k_2 - m_2$ is called a *dynamic vibration absorber* or a *tuned mass damper*.

5.5 Supplementary Examples

Detailed solutions to the following examples are given in (**A**) D. Gross et al. *Formeln und Aufgaben zur Technischen Mechanik 3*, Springer, Berlin 2010, or (**B**) W. Hauger et al. *Aufgaben zur Technischen Mechanik 1-3*, Springer, Berlin 2008.

E5.11

Example 5.11 The system in Fig 5.30 consists of three bars and a beam (with negligible masses) and a block (mass m).

Determine the circular frequency of the free vertical vibrations.

Result: see (**A**)

$$\omega = \frac{1}{l}\sqrt{\frac{1}{ml}\left(6\,EI + \frac{EA\,l^2}{1 + \sqrt{2}}\right)}.$$

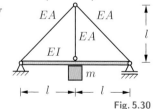

Fig. 5.30

Example 5.12 The system in Fig. 5.31 consists of a homogeneous drum (mass M, radius r), a block (mass m), a spring (spring constant k) and a string (with negligible mass). The support of the drum is frictionless. Assume that there is no slip between the string and the drum.

Determine the natural frequency of the system.

Result: see (**B**) $\omega = \sqrt{\dfrac{2k}{M+2m}}$.

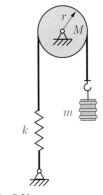

Fig. 5.31

Example 5.13 Two drums rotate in opposite directions as shown in Fig. 5.32. They support a homogeneous board of weight W. The coefficient of kinetic friction between the drums and the board is μ.

Show that the board undergoes a harmonic vibration and determine the natural frequency.

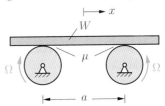

Fig. 5.32

Results: see (**A**) $\ddot{x} + 2\mu\dfrac{g}{a}x = 0$, $\omega = \sqrt{2\mu\dfrac{g}{a}}$.

Example 5.14 A homogeneous bar (weight $W = mg$, length l) is submerged in a viscous fluid and undergoes vibrations about point A (Fig. 5.33). The drag force F_d acting on every point of the bar is proportional to the local velocity (proportionality factor β).

Derive the equation of motion. Assume small amplitudes and neglect the buoyancy. Calculate the value $\beta = \beta^*$ for critical damping.

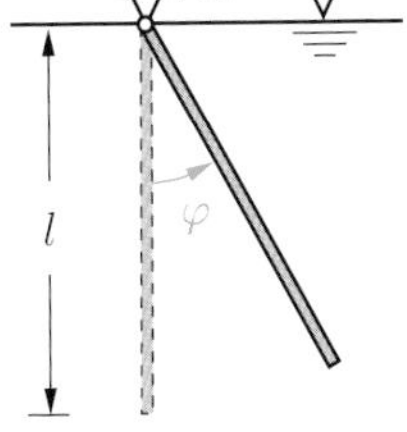

Fig. 5.33

Results: see (**A**) $\ddot{\varphi} + \left(\dfrac{\beta l}{m}\right)\dot{\varphi} + \left(\dfrac{3g}{2l}\right)\varphi = 0$, $\beta^* = \dfrac{m}{l}\sqrt{\dfrac{6\,g}{l}}$.

E5.15

Example 5.15 The pendulum of a clock consists of a homogeneous rod (mass m, length l) and a homogeneous disk (mass M, radius r) whose center is located at a distance a from point A (Fig. 5.34).

Assume small amplitudes and determine the natural frequency of the corresponding oscillations. Choose $m = M$ and $r \ll a$ and calculate the ratio a/l which yields the maximum eigenfrequency.

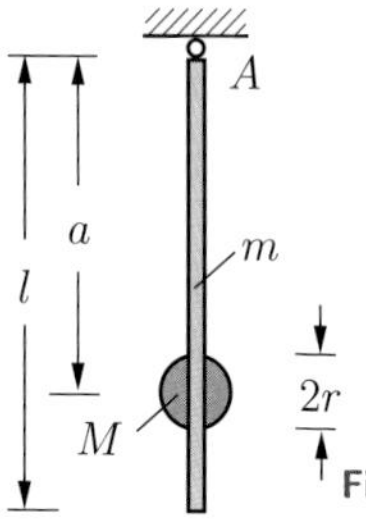

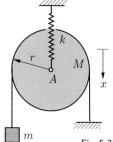

Fig. 5.34

Results: see (**B**) $\omega = \sqrt{\dfrac{(ml + 2Ma)g}{2\Theta_A}}$

with $\Theta_A = m\,l^2/3 + M\,(r^2 + 2a^2)/2$, $\dfrac{a}{l} = \dfrac{1}{2}\left(\sqrt{\dfrac{7}{3}} - 1\right)$.

E5.16

Example 5.16 The system in Fig 5.35 consists of a homogeneous pulley (mass M, radius r), a block (mass m) and a spring (spring constant k).

Determine the equation of motion for the block and its solution for the initial conditions $x(0) = 0$, $v(0) = v_0$. Neglect the mass of the string.

Results: see (**A**)

$$\ddot{x} + \frac{k}{4m + \frac{3}{2}M}\,x = 0 \;\rightarrow\; x(t) = \frac{v_0}{\omega}\sin\omega t\,, \quad \omega = \sqrt{\frac{k}{4m + \frac{3}{2}M}}\,.$$

Fig. 5.35

E5.17

Example 5.17 A wheel (mass m, radius r) rolls without slipping on a vertical circular path (Fig. 5.36). The mass of the rod (length l) can be neglected; the joints are frictionless.

Derive the equation of motion and determine the natural frequency of small oscillations.

Results: see (**B**)

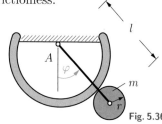

Fig. 5.36

$$\ddot{\varphi} + \frac{2g}{3l}\sin\varphi = 0\,, \quad \omega = \sqrt{\frac{2g}{3l}}\,.$$

Example 5.18 The simple pendulum in Fig. 5.37 is attached to a spring (spring constant k) and a dashpot (damping coefficient d).

a) Determine the maximum value of the damping coefficient d so that the system undergoes vibrations. Assume small amplitudes.

b) Find the damping ratio ζ so that the amplitude is reduced to $1/10$ of its initial value after 10 full cycles. Calculate the corresponding period T_d.

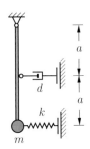

Fig. 5.37

E5.18

Results: see (**A**)

$$d < 8\sqrt{km + \frac{gm^2}{2a}}\,, \quad \zeta = 0.037\,, \quad T_d = 2\pi\sqrt{\frac{2am}{2ak + gm}}\,.$$

Example 5.19 The structure in Fig. 5.38 consists of an elastic beam (flexural rigidity EI, axial rigidity $EA \to \infty$, with negligible mass) and three rigid bars (with negligible masses). The block (mass m) is suspended from a spring (spring constant k).

E5.19

Determine the eigenfrequency of the vertical oscillations of the block.

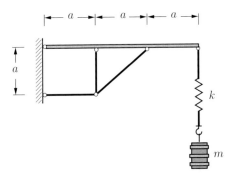

Fig. 5.38

Result: see (**B**) $\omega = \sqrt{\dfrac{k\,EI}{(k\,a^3 + EI)m}}\,.$

E5.20

Example 5.20 A rod (length l, with negligible mass) is elastically supported at point A (Fig. 5.39). The rotational spring (spring constant k_T) is unstretched for $\varphi = 0$. The rod carries a point mass m at its free end.

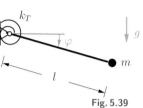

Derive the equation of motion. Determine the spring constant so that $\varphi = \pi/6$ is an equilibrium position. Calculate the natural frequency of small oscillations about this equilibrium position.

Fig. 5.39

Results: see (**B**)

$$\ddot{\varphi} = \frac{g}{l}\cos\varphi - \frac{k_T}{ml^2}\varphi, \quad k_T = \frac{3\sqrt{3}}{\pi}mgl, \quad \omega = \sqrt{\left(\frac{1}{2} + \frac{3\sqrt{3}}{\pi}\right)\frac{g}{l}}.$$

E5.21

Example 5.21 A single storey frame consists of two rigid columns (with masses negligible), a rigid beam of mass m and a spring-dashpot system as shown in Fig. 5.40. The ground is forced to vibration by an earthquake; the acceleration $\ddot{u}_E = b_0\cos\Omega t$ is known from measurements.

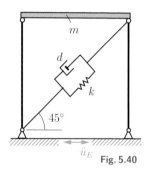

Determine the maximum amplitude of the steady state vibrations. Assume that the system is underdamped and that the vibrations have small amplitudes.

Fig. 5.40

Result: see (**A**) $A = 2\sqrt{2}\,\frac{b_0}{d}\sqrt{\frac{m^3}{k}}$.

E5.22

Example 5.22 The undamped system in Fig. 5.41 consists of a block (mass $m = 4$ kg) and a spring (spring constant $k = 1$ N/m). The block is subjected to a force $F(t)$. The initial conditions $x(0) = x_0 = 1$ m, $\dot{x}(0) = 0$ and the response

$$x(t) = x_0\left[\cos\frac{t}{2t_0} + 20\left(1 - \cos\frac{t-T}{2t_0}\right)\langle t-T\rangle^0\right]$$

to the excitation are given. Here,
$t_0 = 1$ s, $T = 5$ s and $\langle t - T \rangle^0 = 0$
for $t < T$ and $\langle t - T \rangle^0 = 1$ for
$t > T$.

Calculate the force $F(t)$.

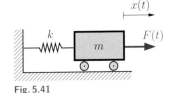

Fig. 5.41

Result: see (**B**) $F(t) = 0$ for $t < T$, $F(t) = 20$ N for $t > T$.

Example 5.23   A simplified model of a car (mass m) is given by
a spring-mass system (Fig. 5.42). The car drives with constant
velocity v_0 over an uneven surface in the form of a sine function
(amplitude U_0, wave length L).

a) Derive the equation of motion and determine the forcing
frequency Ω.

b) Find the amplitude of the vertical vibrations as a function
of the velocity v_0.

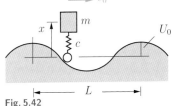

Fig. 5.42

c) Calculate the critical velocity v_c (resonance!).

Results: see (**A**) $m\ddot{x} + kx = kU_0\cos\Omega t$,

$$\Omega = \frac{2\pi v_0}{L}, \qquad x_0 = \frac{U_0}{1 - \Omega^2 \frac{m}{k}}, \qquad v_c = \frac{L}{2\pi}\sqrt{\frac{k}{m}}.$$

Example 5.24   A homogeneous wheel (mass m) is attached to a
spring (spring constant k). The
wheel rolls without slipping on
a rough surface which moves
according to the function $u =
u_0\cos\Omega t$ (Fig. 5.43).

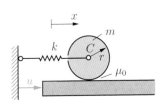

a) Determine the amplitude of
the steady state vibrations.

Fig. 5.43

b) Calculate the coefficient μ_0 of static friction which is necessary
to prevent slipping.

Results: see (**A**) $|x_0| = \dfrac{u_0}{3\left|\dfrac{2}{3}\dfrac{k}{m\Omega^2} - 1\right|}$, $\mu_0 \geq \dfrac{u_0\Omega^2}{3g}\left|\dfrac{\dfrac{k}{m\Omega^2} - 1}{\dfrac{2}{3}\dfrac{k}{m\Omega^2} - 1}\right|$.

E5.23

E5.24

E5.25

Example 5.25 A small homogeneous disk (mass m, radius r) is attached to a large homogeneous disk (mass M, radius R) as shown in Fig. 5.44. The torsion spring (spring constant k_T) is unstretched in the position shown.

Determine the eigenfrequency of the vertical oscillations. Assume small amplitudes.

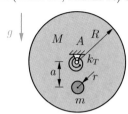

Fig. 5.44

Result: see (**A**) $\omega = \sqrt{\dfrac{k_T + mga}{\dfrac{1}{2}MR^2 + m(\dfrac{r^2}{2} + a^2)}}$

E5.26

Example 5.26 The systems ① and ② in Fig. 5.45 consist of two beams (masses negligible, flexural rigidity EI), a spring (spring constant k) and a box (mass m).

Determine the spring constants k^* of the equivalent springs for the two systems.

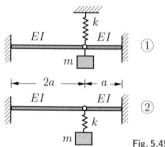

Fig. 5.45

Results: see (**A**)   ①: $k^* = \dfrac{27EI + 8ka^3}{8a^3}$,   ②: $k^* = \dfrac{27EI}{8a^3 + 27\dfrac{EI}{k}}$

E5.27

Example 5.27 A homogeneous wheel (mass m, moment of inertia Θ_C, radius r) rolls without slipping on a rough beam (mass M). The beam moves without friction on roller supports (Fig. 5.46).

Determine the natural frequency of the system.

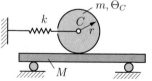

Fig. 5.46

Result: see (**A**) $\omega = \sqrt{\dfrac{k}{m + \dfrac{M}{1 + Mr^2/\Theta_C}}}$

5.6 Summary

- Differential equation for harmonic vibrations:

$$\ddot{x} + \omega^2 x = 0 \quad \rightarrow \quad x = C\cos(\omega t - \alpha)\,,$$
$$\omega = 2\pi/T = 2\pi f \quad \text{circular frequency.}$$

- Springs in parallel, springs in series:

$$k^* = \sum k_j\,, \quad \frac{1}{k^*} = \sum \frac{1}{k_j}\,.$$

- Underdamped free vibrations:

$$\ddot{x} + 2\xi\,\dot{x} + \omega^2 x = 0 \quad \rightarrow \quad x = C\,\mathrm{e}^{-\xi t}\cos(\omega_d\,t - \alpha)\,,$$
$$\xi \quad \text{normalized damping coefficient,} \quad \omega_d = \omega\sqrt{1-\zeta^2}$$
circular frequency, $\quad \zeta = \xi/\omega \quad$ damping factor.

- Undamped forced vibrations:

$$\ddot{x} + \omega^2 x = \omega^2 x_0 \cos\Omega t \quad \rightarrow \quad x_p = x_0\,V\cos\Omega t\,,$$
$$\Omega \quad \text{forcing frequency,} \quad V = \frac{1}{1-\eta^2} \quad \text{magnification factor,}$$
$$\eta = \Omega/\omega \quad \text{frequency ratio,} \quad \text{resonance:} \quad \Omega = \omega\,, \quad V \to \infty\,.$$

- The magnification factor V depends on the type of the forcing (e.g., excitation through a force, rotating eccentric mass) in the case of damped forced vibrations.
  The phase angle φ is independent of the type of forcing.
- A system with two degrees of freedom has two natural frequencies: ω_1 and ω_2.
- First principal mode of vibration: both masses vibrate in phase with the frequency ω_1. Second principal mode of vibration: both masses vibrate $180°$ out of phase with the frequency ω_2. Exception: degenerate systems.
- Dynamic vibration absorber (tuned mass damper): for a given mass m_2 and a given spring constant k_2 a vibration absorption takes place for a forcing frequency of $\Omega_T = \sqrt{k_2/m_2}$.

Chapter 6

Non-Inertial Reference Frames

6 Non-Inertial Reference Frames

———— Objectives: According to Section 1.2.1 Newton's law in the form $m\,\boldsymbol{a} = \boldsymbol{F}$ is valid in a reference frame that is *fixed* in space. Such a reference frame is an inertial frame. The notion of an inertial frame will be discussed in more detail in Section 6.2.

Sometimes, however, it is advantageous to describe the motion of a body relative to a *moving* frame of reference. For this reason we need to know the relationships between the kinematic quantities in fixed and in moving frames and we have to apply Newton's law in a form that is valid in a moving reference frame.

6.1 Kinematics of Relative Motion

6.1

6.1.1 Translating Reference Frames

In this chapter we will investigate the spatial motion of a particle P using two different coordinate systems (Fig. 6.1). The x, y, z-system is fixed in space. The ξ, η, ζ-system with the unit vectors e_ξ, e_η and e_ζ undergoes a *translation* relative to the fixed system (rotating systems will be considered in Section 6.1.2).

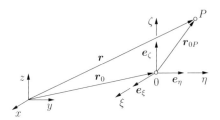

Fig. 6.1

The position vector r to point P can be written as

$$r = r_0 + r_{0P} \tag{6.1}$$

where $r_{0P} = \xi \, e_\xi + \eta \, e_\eta + \zeta \, e_\zeta$. We call the velocity of point P measured in the fixed system the *absolute velocity*. It is obtained by taking the time derivative of the position vector (cf. Section 1.1.1):

$$v_a = \dot{r} = \dot{r}_0 + \dot{r}_{0P} \tag{6.2}$$

where $\dot{r}_{0P} = \dot{\xi} \, e_\xi + \dot{\eta} \, e_\eta + \dot{\zeta} \, e_\zeta$ (the unit vectors have a constant direction and are therefore independent of the time). We added the subscript a at v_a in order to distinguish the absolute velocity from other velocities which will be introduced in the following.

Analogously, the acceleration of point P measured in the fixed coordinate system is called the *absolute acceleration*. It is defined as the time derivative of the absolute velocity:

$$a_a = \dot{v}_a = \ddot{r}_0 + \ddot{r}_{0P} \tag{6.3}$$

with $\ddot{r}_{0P} = \ddot{\xi} \, e_\xi + \ddot{\eta} \, e_\eta + \ddot{\zeta} \, e_\zeta$.

The terms $\dot{\boldsymbol{r}}_0$ and $\ddot{\boldsymbol{r}}_0$, respectively, in (6.2) and (6.3) are the absolute velocity and the absolute acceleration of the origin 0 of the translating ξ, η, ζ-system. We refer to $\dot{\boldsymbol{r}}_0 = \boldsymbol{v}_f$ and $\ddot{\boldsymbol{r}}_0 = \boldsymbol{a}_f$ as the *velocity of reference frame* and the *acceleration of reference frame*. The terms $\dot{\boldsymbol{r}}_{0P}$ and $\ddot{\boldsymbol{r}}_{0P}$, respectively, are the velocity and the acceleration of point P relative to the translating frame. Therefore, $\dot{\boldsymbol{r}}_{0P} = \boldsymbol{v}_r$ is called the *relative velocity* and $\ddot{\boldsymbol{r}}_{0P} = \boldsymbol{a}_r$ is referred to as the *relative acceleration* of point P. They are the velocity and the acceleration which are measured by an observer who is fixed in the moving frame. Thus, Equations (6.2) and (6.3) can be written as

$$\boldsymbol{v}_a = \boldsymbol{v}_f + \boldsymbol{v}_r, \quad \boldsymbol{a}_a = \boldsymbol{a}_f + \boldsymbol{a}_r \tag{6.4}$$

in the case of a translating frame: the absolute velocity (acceleration) is the vector sum of the velocity (acceleration) of reference frame and the relative velocity (acceleration).

6.1.2 Translating and Rotating Reference Frames
We will now investigate the velocity and the acceleration, if the moving frame undergoes a translation *and* a rotation relative to the fixed frame.

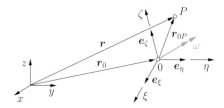

Fig. 6.2

The position vector $\boldsymbol{r}$ to point P is again given by (Fig. 6.2)

$$\boldsymbol{r} = \boldsymbol{r}_0 + \boldsymbol{r}_{0P} \tag{6.5}$$

where $\boldsymbol{r}_{0P} = \xi \, \boldsymbol{e}_\xi + \eta \, \boldsymbol{e}_\eta + \zeta \, \boldsymbol{e}_\zeta$. The *absolute velocity* of point P is obtained by taking the time derivative of the position vector:

$$\boldsymbol{v}_a = \dot{\boldsymbol{r}} = \dot{\boldsymbol{r}}_0 + \dot{\boldsymbol{r}}_{0P}. \tag{6.6}$$

Since the directions of the unit vectors $\boldsymbol{e}_\xi, \boldsymbol{e}_\eta$ and $\boldsymbol{e}_\zeta$ are not constant in a rotating system we have

$$\dot{\boldsymbol{r}}_{0P} = (\dot{\xi}\,\boldsymbol{e}_\xi + \dot{\eta}\,\boldsymbol{e}_\eta + \dot{\zeta}\,\boldsymbol{e}_\zeta) + (\xi\,\dot{\boldsymbol{e}}_\xi + \eta\,\dot{\boldsymbol{e}}_\eta + \zeta\,\dot{\boldsymbol{e}}_\zeta). \tag{6.7}$$

The moving frame rotates with the angular velocity $\boldsymbol{\omega}$. Therefore, in analogy to (3.5), the time derivatives of the unit vectors are given by

$$\dot{\boldsymbol{e}}_\xi = \boldsymbol{\omega} \times \boldsymbol{e}_\xi, \quad \dot{\boldsymbol{e}}_\eta = \boldsymbol{\omega} \times \boldsymbol{e}_\eta, \quad \dot{\boldsymbol{e}}_\zeta = \boldsymbol{\omega} \times \boldsymbol{e}_\zeta. \tag{6.8}$$

Thus,

$$\begin{aligned}
\xi\,\dot{\boldsymbol{e}}_\xi + \eta\,\dot{\boldsymbol{e}}_\eta + \zeta\,\dot{\boldsymbol{e}}_\zeta &= \xi\,\boldsymbol{\omega} \times \boldsymbol{e}_\xi + \eta\,\boldsymbol{\omega} \times \boldsymbol{e}_\eta + \zeta\,\boldsymbol{\omega} \times \boldsymbol{e}_\zeta \\
&= \boldsymbol{\omega} \times (\xi\,\boldsymbol{e}_\xi + \eta\,\boldsymbol{e}_\eta + \zeta\,\boldsymbol{e}_\zeta) = \boldsymbol{\omega} \times \boldsymbol{r}_{0P},
\end{aligned}$$

and (6.7) becomes

$$\dot{\boldsymbol{r}}_{0P} = \frac{\mathrm{d}\boldsymbol{r}_{0P}}{\mathrm{d}t} = (\dot{\xi}\,\boldsymbol{e}_\xi + \dot{\eta}\,\boldsymbol{e}_\eta + \dot{\zeta}\,\boldsymbol{e}_\zeta) + \boldsymbol{\omega} \times \boldsymbol{r}_{0P}. \tag{6.9}$$

The terms in the parentheses represent the time derivative of $\boldsymbol{r}_{0P}$ relative to the *moving* frame. We distinguish the time derivatives relative to the moving frame by an asterisk:

$$\frac{\mathrm{d}^*\boldsymbol{r}_{0P}}{\mathrm{d}t} = \dot{\xi}\,\boldsymbol{e}_\xi + \dot{\eta}\,\boldsymbol{e}_\eta + \dot{\zeta}\,\boldsymbol{e}_\zeta.$$

Then (6.9) can be written as

$$\dot{\boldsymbol{r}}_{0P} = \frac{\mathrm{d}^*\boldsymbol{r}_{0P}}{\mathrm{d}t} + \boldsymbol{\omega} \times \boldsymbol{r}_{0P}. \tag{6.10}$$

This relationship between the time derivative of vector $\boldsymbol{r}_{0P}$ relative to the fixed system and to the moving system, respectively, holds analogously for arbitrary vectors.

Inserting (6.10) and the velocity $\boldsymbol{v}_0$ of the origin of the moving system into (6.6) yields

$$v_a = v_0 + \boldsymbol{\omega} \times \boldsymbol{r}_{0P} + \frac{\mathrm{d}^* \boldsymbol{r}_{0P}}{\mathrm{d}t} \,. \tag{6.11}$$

This equation may be written compactly as

$$\boldsymbol{v}_a = \boldsymbol{v}_f + \boldsymbol{v}_r \tag{6.12a}$$

where

$$\boldsymbol{v}_f = \boldsymbol{v}_0 + \boldsymbol{\omega} \times \boldsymbol{r}_{0P} \,,$$
$$\boldsymbol{v}_r = \frac{\mathrm{d}^* \boldsymbol{r}_{0P}}{\mathrm{d}t} \,. \tag{6.12b}$$

The *velocity* $\boldsymbol{v}_f$ is the velocity which point P would have if it was fixed in the moving frame. The *relative velocity* $\boldsymbol{v}_r$ is the velocity of point P relative to the moving system. It is the velocity which is measured by an observer fixed in the moving frame.

The *absolute acceleration* of P is obtained by taking the time derivative of the absolute velocity:

$$\boldsymbol{a}_a = \dot{\boldsymbol{v}}_a = \dot{\boldsymbol{v}}_f + \dot{\boldsymbol{v}}_r = \dot{\boldsymbol{v}}_0 + (\boldsymbol{\omega} \times \boldsymbol{r}_{0P})^{\boldsymbol{\cdot}} + \dot{\boldsymbol{v}}_r \,. \tag{6.13}$$

Using (6.10) and (6.12b), the second term on the right-hand side of (6.13) can be written as

$$(\boldsymbol{\omega} \times \boldsymbol{r}_{0P})^{\boldsymbol{\cdot}} = \dot{\boldsymbol{\omega}} \times \boldsymbol{r}_{0P} + \boldsymbol{\omega} \times \dot{\boldsymbol{r}}_{0P}$$
$$= \dot{\boldsymbol{\omega}} \times \boldsymbol{r}_{0P} + \boldsymbol{\omega} \times \left(\frac{\mathrm{d}^* \boldsymbol{r}_{0P}}{\mathrm{d}t} + \boldsymbol{\omega} \times \boldsymbol{r}_{0P} \right) \tag{6.14}$$
$$= \dot{\boldsymbol{\omega}} \times \boldsymbol{r}_{0P} + \boldsymbol{\omega} \times \boldsymbol{v}_r + \boldsymbol{\omega} \times (\boldsymbol{\omega} \times \boldsymbol{r}_{0P}) \,.$$

In analogy to (6.10) the last term in (6.13) is given by

$$\dot{\boldsymbol{v}}_r = \frac{\mathrm{d}^* \boldsymbol{v}_r}{\mathrm{d}t} + \boldsymbol{\omega} \times \boldsymbol{v}_r \tag{6.15}$$

where

$$\frac{\mathrm{d}^* \boldsymbol{v}_r}{\mathrm{d}t} = \ddot{\xi} \boldsymbol{e}_\xi + \ddot{\eta} \boldsymbol{e}_\eta + \ddot{\zeta} \boldsymbol{e}_\zeta \,.$$

If we insert (6.14) and (6.15) into (6.13) we obtain

$$a_a = \dot{v}_0 + \dot{\omega} \times r_{0P} + \omega \times (\omega \times r_{0P}) + \frac{d^* v_r}{dt} + 2\,\omega \times v_r . \quad (6.16)$$

Here, $\dot{v}_0 = a_0$ is the acceleration of point 0. We can write (6.16) in the form

$$a_a = a_f + a_r + a_c \qquad (6.17a)$$

where

$$a_f = a_0 + \dot{\omega} \times r_{0P} + \omega \times (\omega \times r_{0P}),$$

$$a_r = \frac{d^* v_r}{dt} = \frac{d^{2*} r_{0P}}{dt^2}, \qquad (6.17b)$$

$$a_c = 2\,\omega \times v_r .$$

The *acceleration* a_f is the acceleration which point P would have if it was fixed in the moving frame (compare (3.8)). The *relative acceleration* a_r is the acceleration of point P relative to the moving system. It is the acceleration which is measured by an observer fixed in the moving frame. The term a_c in (6.17a,b) is called the *Coriolis acceleration* after Gaspard Gustave de Coriolis (1792–1843). It is orthogonal to ω and to v_r and it vanishes if a) $\omega = 0$, b) $v_r = 0$ or c) v_r is parallel to ω.

The relations for v_f and a_f in (6.12b) and (6.17b) can be simplified with the aid of polar coordinates in the special case of a planar motion. We choose the coordinate system in such a way that the axes x, y and ξ, η lie in the plane of the motion (Fig. 6.3). Then the angular velocity vector ω points in the direction of the ζ-axis. With $r_{0P} = r e_r$ and $\omega = \omega e_\zeta$ we obtain

$$\omega \times r_{0P} = r\omega e_\varphi, \quad \dot{\omega} \times r_{0P} = r\dot{\omega} e_\varphi,$$

$$\omega \times (\omega \times r_{0P}) = -r\omega^2 e_r .$$

Thus, the relations for the fixed frame velocity and the fixed frame acceleration reduce to

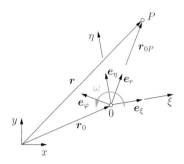

Fig. 6.3

$$\boldsymbol{v}_f = \boldsymbol{v}_0 + r\omega\boldsymbol{e}_\varphi, \quad \boldsymbol{a}_f = \boldsymbol{a}_0 + r\dot\omega\boldsymbol{e}_\varphi - r\omega^2\boldsymbol{e}_r \tag{6.18}$$

(compare Section 3.1.3).

E6.1   **Example 6.1** Two circular disks (radii $R_1 = 2R$, $R_2 = R$) are supported as shown in Fig. 6.4a. They rotate without relative slipping. The constant angular velocity ω_1 of disk ① is given. An observer is located at point 0; he rotates with disk ①.

Determine the velocity and the acceleration of point P on disk ② as measured by the observer.

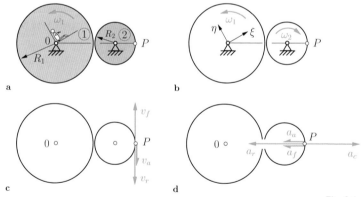

Fig. 6.4

Solution Since there is no slip between the disks, the points of contact on both disks have the same velocity. Thus,

$$R_1\,\omega_1 = R_2\,\omega_2 \quad \rightarrow \quad 2\,R\omega_1 = R\omega_2 \quad \rightarrow \quad \omega_2 = 2\,\omega_1\,.$$

The rotating coordinate system ξ, η (Fig. 6.4b) is fixed to disk ①. The rotating observer measures the relative velocity

$$\boldsymbol{v}_r = \boldsymbol{v}_a - \boldsymbol{v}_f \tag{a}$$

of point P (see (6.12a,b)). The absolute velocity of point P is given by $v_a = R_2\,\omega_2 = 2\,R\,\omega_1$ (directed downwards, see Fig. 6.4c) since disk ② rotates with the angular velocity $\omega_2 = 2\,\omega_1$. In order to determine the velocity of the frame of reference $\boldsymbol{v}_f$ we imagine point P to be fixed in the rotating ξ, η-system, i.e. fixed in the rotating disk ①. In this case it would move on a circle with the radius $4\,R$ and angular velocity ω_1. Thus, we get $v_f = 4\,R\omega_1$ (directed upwards), and (a) yields

$$\underline{\underline{v_r = v_a - v_f = 2\,R\omega_1 - (-\,4\,R\omega_1) = 6\,R\omega_1}}\,. \tag{b}$$

The relative velocity v_r is directed downwards (Fig. 6.4c).

The rotating observer measures the relative acceleration of point P. According to (6.17a,b) it is given by

$$\boldsymbol{a}_r = \boldsymbol{a}_a - \boldsymbol{a}_f - \boldsymbol{a}_c\,. \tag{c}$$

The absolute acceleration is $a_a = R_2\,\omega_2^2 = 4\,R\omega_1^2$ (centripetal acceleration); it points to the left (Fig. 6.4d). To determine the acceleration of the frame of reference we again imagine point P to be fixed in disk ①. Then we get $a_f = 4\,R\omega_1^2$ (directed to the left). The Coriolis acceleration is found from $\boldsymbol{a}_c = 2\,\boldsymbol{\omega}_1 \times \boldsymbol{v}_r$ (vector $\boldsymbol{\omega}_1$ is orthogonal to the plane of the disks). We obtain $a_c = 2\,\omega_1\,6\,R\omega_1 = 12\,R\omega_1^2$ (to the right). Thus, Equation (c) yields

$$\underline{\underline{a_r = a_a - a_f - a_c = 12\,R\omega_1^2}}\,.$$

The relative acceleration points to the left (Fig. 6.4d).

6.2 Kinetics of Relative Motion

According to Section 1.2.1 Newton's 2nd Law in a reference frame fixed in space can be stated as mass × absolute acceleration = force. If we insert the absolute acceleration given in (6.17a) we obtain

$$ma_a = F \quad \rightarrow \quad m\left(a_f + a_r + a_c\right) = F\,.$$

Solving for the relative acceleration yields the equation of motion relative to a moving reference frame:

$$ma_r = F - ma_f - ma_c\,.$$

In addition to the applied (real) forces F we have the terms $-ma_f$ and $-ma_c$ on the right-hand side. We now introduce *d'Alembert's inertial force* F_f and the *Coriolis force* F_c:

$$F_f = -ma_f, \quad F_c = -ma_c\,. \tag{6.19}$$

Then the equation of motion can be written in the form

$$ma_r = F + F_f + F_c\,. \tag{6.20}$$

Thus, d'Alembert's inertial force F_f and the Coriolis force F_c have to be added to the real forces F in the case of a moving reference frame. The forces F_f and F_c are *fictitious (pseudo) forces*, compare Section 4.1. If the reference frame is translating ($\omega = 0$), then the Coriolis force in (6.20) vanishes.

Let us now consider the special case of a reference frame that is *translating with a constant velocity* (uniform motion). Then the acceleration a_0 and the angular velocity ω are equal to zero. Hence, the fixed frame acceleration and the Coriolis acceleration vanish according to (6.17b) and the corresponding fictitious forces vanish according to (6.19). In this case the relative acceleration coincides with the absolute acceleration ($a_r = a_a$). Then the equation of motion (6.20) reduces to

$$ma_r = F\,,$$

which is identical with the equation of motion in a fixed reference frame.

All reference frames in which the equation of motion takes the form $m\boldsymbol{a}_r = \boldsymbol{F}$ are called *inertial frames*. Accordingly, fixed as well as uniformly moving frames are inertial frames. There are no fictitious forces if one describes the motion in such a frame.

Example 6.2 Point 0 of a simple pendulum (mass m, length l) moves upwards with a constant acceleration a_0 (Fig. 6.5a).

E6.2

Derive the equation of motion. Determine the natural frequency of the vibration (assume small amplitudes) and the force in the string.

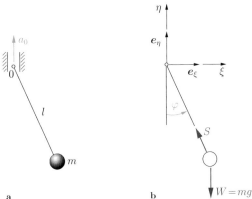

Fig. 6.5 a b $W = mg$

Solution We introduce the ξ, η-coordinate system as depicted in Fig. 6.5b. It is a *translating* coordinate system with point 0 as the origin. According to (6.20) the equation of motion in the *moving* system is

$$m\boldsymbol{a}_r = \boldsymbol{F} + \boldsymbol{F}_f + \boldsymbol{F}_c . \tag{a}$$

The (real) force $\boldsymbol{F}$ acting at the mass is given by (see Fig. 6.5b)

$$\boldsymbol{F} = -S \sin \varphi \, \boldsymbol{e}_\xi + (S \cos \varphi - mg)\boldsymbol{e}_\eta \tag{b}$$

and the fictitious forces are

$$\boldsymbol{F}_f = -m\boldsymbol{a}_f = -ma_0\boldsymbol{e}_\eta, \quad \boldsymbol{F}_c = -m\boldsymbol{a}_c = \boldsymbol{0} , \tag{c}$$

since $\boldsymbol{\omega} = \boldsymbol{0}$. The components of the relative acceleration $\boldsymbol{a}_r$ follow from the coordinates of the point mass in the moving system through differentiation:

$$\xi = l\sin\varphi, \qquad\qquad \eta = -l\cos\varphi,$$

$$\dot\xi = l\dot\varphi\cos\varphi, \qquad\qquad \dot\eta = l\dot\varphi\sin\varphi,$$

$$\ddot\xi = l\ddot\varphi\cos\varphi - l\dot\varphi^2\sin\varphi, \qquad \ddot\eta = l\ddot\varphi\sin\varphi + l\dot\varphi^2\cos\varphi.$$

This yields the relative acceleration

$$\boldsymbol{a}_r = \ddot\xi\,\boldsymbol{e}_\xi + \ddot\eta\,\boldsymbol{e}_\eta = (l\ddot\varphi\cos\varphi - l\dot\varphi^2\sin\varphi)\boldsymbol{e}_\xi$$
$$+ (l\ddot\varphi\sin\varphi + l\dot\varphi^2\cos\varphi)\boldsymbol{e}_\eta. \tag{d}$$

If we insert (b) - (d) into (a) we obtain the components of the equation of motion in the directions of the axes ξ and η:

$$m\,(l\ddot\varphi\cos\varphi - l\dot\varphi^2\sin\varphi) = -S\sin\varphi, \tag{e}$$

$$m\,(l\ddot\varphi\sin\varphi + l\dot\varphi^2\cos\varphi) = S\cos\varphi - mg - ma_0. \tag{f}$$

These are two equations for the unknowns φ and S. We can eliminate S if we multiply (e) with $\cos\varphi$ and (f) with $\sin\varphi$ and subsequently add the two equations. This leads to the equation of motion

$$ml\ddot\varphi = -mg\sin\varphi - ma_0\sin\varphi \quad\rightarrow\quad \underline{\underline{\ddot\varphi + \frac{g+a_0}{l}\sin\varphi = 0}}. \tag{g}$$

In the case of small amplitudes ($\sin\varphi \approx \varphi$) of vibration, Equation (g) reduces to the differential equation for harmonic oscillations

$$\ddot\varphi + \omega^2\varphi = 0$$

with the natural frequency

$$\underline{\underline{\omega = \sqrt{\frac{g+a_0}{l}}}}.$$

If point 0 is accelerated upwards, the natural frequency is larger than the natural frequency for a fixed point 0. On the other hand, if point 0 is accelerated downwards ($a_0 < 0$) the pendulum oscillates more slowly. In the special case of point 0 in a free fall ($a_0 = -g$) we get $\omega = 0$.

If we multiply (e) with $\sin\varphi$ and (f) with $\cos\varphi$ and subtract the equations we obtain the force in the string:

$$\underline{S = m\left[l\dot{\varphi}^2 + (g + a_0)\cos\varphi\right]}.$$

Example 6.3   A point mass m moves in a frictionless slot in a disk (Fig. 6.6a). It is supported by two identical springs (spring constant of each spring $k/2$). The disk rotates with the constant angular velocity ω about point A.

E6.3

Describe the motion of the mass relative to the rotating disk. Determine the force that is exerted from the slot on the mass. Neglect the weight of the mass.

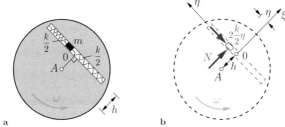

Fig. 6.6   a   b

Solution   The mass can only move in the slot: it undergoes a rectilinear motion relative to the disk. We introduce the rotating coordinate system ξ, η shown in Fig. 6.6b. Then the motion takes place in the direction of the η-axis. Hence, $\xi = 0$, $\dot{\xi} = 0$, $\ddot{\xi} = 0$.

The equation of motion (6.20) in the rotating system is given by

$$m\boldsymbol{a}_r = \boldsymbol{F} + \boldsymbol{F}_f + \boldsymbol{F}_c. \tag{a}$$

The forces acting on the mass are the contact force N and the force $2\frac{k}{2}\eta$ exerted by the springs (see Fig. 6.6b). Therefore the external force follows as

$$\boldsymbol{F} = N\boldsymbol{e}_\xi - k\,\eta\,\boldsymbol{e}_\eta. \tag{b}$$

In order to determine the fictitious forces $\boldsymbol{F}_f$ and $\boldsymbol{F}_c$ we first have to find the acceleration of the frame of reference and the Coriolis acceleration. The origin 0 of the rotating system is located at a distance h from the center A of the disk. Its acceleration is therefore given by $\boldsymbol{a}_0 = -h\omega^2\boldsymbol{e}_\xi$ (circular motion with $\dot{\omega} = 0$). With $r = \eta$ and $\boldsymbol{e}_r = \boldsymbol{e}_\eta$ we obtain from (6.18)

$$a_f = -h\omega^2 e_\xi - \eta\omega^2 e_\eta \quad \rightarrow \quad F_f = m(h\omega^2 e_\xi + \eta\omega^2 e_\eta). \quad \text{(c)}$$

The vector of the angular velocity $\boldsymbol{\omega}$ is orthogonal to the ξ, η-plane. Therefore, the Coriolis acceleration and the Coriolis force follow from (6.17b) with $v_r = \dot{\eta}e_\eta$ as

$$a_c = -2\,\dot{\eta}\omega\,e_\xi \quad \rightarrow \quad F_c = 2\,m\dot{\eta}\omega\,e_\xi. \quad \text{(d)}$$

Inserting the relative acceleration

$$a_r = \ddot{\eta}\,e_\eta$$

and the forces (b) - (d) into (a) leads to

$$
\begin{aligned}
0 &= N + mh\omega^2 + 2\,m\dot{\eta}\omega, \\
m\ddot{\eta} &= -k\,\eta + m\eta\omega^2.
\end{aligned}
\quad \text{(e)}
$$

The second equation yields

$$\ddot{\eta} + \left(\frac{k}{m} - \omega^2\right)\eta = 0.$$

Thus, the point mass undergoes a harmonic vibration $\eta = A\cos\omega^* t + B\sin\omega^* t$ relative to the rotating disk if $\omega^2 < k/m$. The circular frequency $\omega^* = \sqrt{k/m - \omega^2}$ is smaller than the circular frequency $\sqrt{k/m}$ in the case of a non-rotating disk ($\omega = 0$).

The first equation in (e) yields the contact force

$$\underline{\underline{N = -m\left(h\omega^2 + 2\,\dot{\eta}\omega\right).}}$$

Example 6.4 A point mass ($m = 1000$ kg) moves on a great circle on the rotating earth ($R = 6370$ km) towards the north pole (Fig. 6.7a). It has a relative velocity $v_r = 100$ km/h.

Determine the maximum d'Alembert's inertial force and the maximum Coriolis force. Neglect the motion of the earth around the sun.

Solution The earth performs one revolution about its north-south axis in 24 hours. Thus, the angular velocity of its rotation is given by

$$\omega = \frac{2\,\pi}{24 \cdot 3600} = 7.27 \cdot 10^{-5}\,\mathrm{s}^{-1}.$$

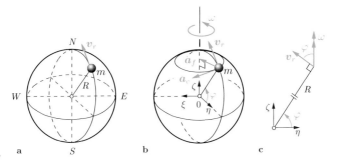

Fig. 6.7

We introduce the rotating ξ, η, ζ-coordinate system which is fixed to the earth (Fig. 6.7b). Then the angular velocity vector is $\boldsymbol{\omega} = \omega \boldsymbol{e}_\zeta$.

The fixed frame acceleration is equal to the acceleration of the point of the earth at the instantaneous location of the point mass. It is obtained from (6.17b) with $\boldsymbol{a}_0 = \boldsymbol{0}$ and $\dot{\boldsymbol{\omega}} = \boldsymbol{0}$:

$$\boldsymbol{a}_f = \boldsymbol{\omega} \times (\boldsymbol{\omega} \times \boldsymbol{r}_{0P}) = -R\cos\varphi\,\omega^2 \boldsymbol{e}_\eta\,.$$

It is orthogonal to the rotation axis of the earth. The Coriolis acceleration follows from (6.17a,b):

$$\boldsymbol{a}_c = 2\,\boldsymbol{\omega} \times \boldsymbol{v}_r = 2\,\omega v_r \sin\varphi\,\boldsymbol{e}_\xi$$

(compare Fig. 6.7c). It is tangential to the small circle and points westwards. D'Alembert's inertial force and the Coriolis force are thus

$$\boldsymbol{F}_f = mR\omega^2\cos\varphi\,\boldsymbol{e}_\eta,\quad \boldsymbol{F}_c = -\,2\,m\omega v_r \sin\varphi\,\boldsymbol{e}_\xi\,.$$

D'Alembert's inertial force reaches its maximum value at the equator ($\varphi = 0$):

$$\underline{\underline{F_{f\,\text{max}} = mR\omega^2 \approx 34\text{ N}\,.}}$$

It is small as compared with the weight of the point mass ($W = mg \approx 10^4$ N). The Coriolis force has its maximum value at the North Pole ($\varphi = \pi/2$):

$$F_{c\,\mathrm{max}} = 2\,m\omega v_r \approx 4\ \mathrm{N}\,.$$

A force that opposes the Coriolis force has to act on the mass. Otherwise the point mass would deviate from the great circle.

6.3 Supplementary Examples

Detailed solutions to the following examples are given in (**A**) D. Gross et al. *Formeln und Aufgaben zur Technischen Mechanik 3*, Springer, Berlin 2010, or (**B**) W. Hauger et al. *Aufgaben zur Technischen Mechanik 1-3*, Springer, Berlin 2008.

Example 6.5 Point A of the simple pendulum (mass m, length l) in Fig. 6.8 moves with a constant acceleration a_0 to the right.

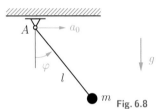

   Derive the equation of motion.

Result: see (**A** and **B**) $l\ddot{\varphi} + g\sin\varphi + a_0\cos\varphi = 0\,.$

Example 6.6 The two disks in Fig. 6.9 rotate with constant angular velocities Ω and ω about their respective axes.

   Determine the absolute acceleration of point P at the instant shown.

Result: see (**B**)

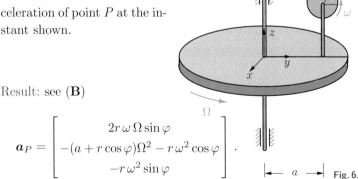

$$\boldsymbol{a}_P = \begin{bmatrix} 2r\,\omega\,\Omega\sin\varphi \\ -(a+r\cos\varphi)\Omega^2 - r\,\omega^2\cos\varphi \\ -r\,\omega^2\sin\varphi \end{bmatrix}\,.$$

Example 6.7 A horizontal circular platform (radius r) rotates with constant angular velocity Ω (Fig. 6.10). A block (mass m) is locked in a frictionless slot at a distance a from the center of the platform. At time $t = 0$ the block is released.

E6.7

Determine the velocity v_r of the block relative to the platform when it reaches the rim of the platform.

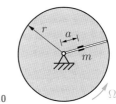

Result: see (**B**)　$v_r = \Omega\sqrt{r^2 - a^2}$. **Fig. 6.10**

Example 6.8 A simple pendulum is attached to point 0 of a circular disk (Fig. 6.11). The disk rotates with a constant angular velocity Ω; the pendulum oscillates in the horizontal plane.

E6.8

Determine the circular frequency of the oscillations. Assume small amplitudes and neglect the weight of the mass.

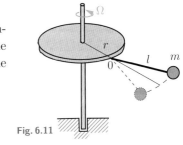

Result: see (**A**)　$\omega = \sqrt{\dfrac{r}{l}}\,\Omega$. **Fig. 6.11**

Example 6.9 A drum rotates with angular velocity ω about point B (Fig. 6.12). Pin C is fixed to the drum; it moves in the slot of link AD.

E6.9

Determine the angular velocity ω_{AD} of link AD and the velocity v_r of the pin relative to the link at the instant shown.

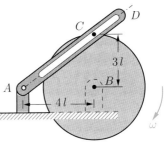

Results: see (**B**)
$$\omega_{AD} = \frac{9\,\omega}{25}, \qquad v_r = \frac{12\omega\,l}{5}.$$
Fig. 6.12

E6.10

Example 6.10 A point P moves along a circular path (radius r) on a platform with a constant relative velocity v_r (Fig. 6.13). The platform rotates with a constant angular velocity ω about point A. The eccentricity e is given.

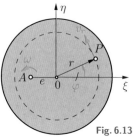

Determine the relative, fixed frame-, Coriolis, and absolute accelerations of P.

Fig. 6.13

Results: see (**A**) $\boldsymbol{a}_r = -\dfrac{v_r^2}{r}(\boldsymbol{e}_\xi \cos\varphi + \boldsymbol{e}_\eta \sin\varphi)$,

$$\boldsymbol{a}_f = -(e + r\cos\varphi)\omega^2 \boldsymbol{e}_\xi - r\omega^2 \sin\varphi\, \boldsymbol{e}_\eta,$$

$$\boldsymbol{a}_c = -2\omega v_r(\boldsymbol{e}_\xi \cos\varphi + \boldsymbol{e}_\eta \sin\varphi),$$

$$\boldsymbol{a} = \boldsymbol{a}_f + \boldsymbol{a}_r + \boldsymbol{a}_c = -[e\omega^2 + r\left(\omega + \frac{v_r}{r}\right)^2 \cos\varphi]\boldsymbol{e}_\xi$$
$$-r\left(\omega + \frac{v_r}{r}\right)^2 \sin\varphi\, \boldsymbol{e}_\eta.$$

E6.11

Example 6.11 A circular ring (radius r) rotates with constant angular velocity Ω about the x-axis (Fig. 6.14). A point mass m moves without friction inside the ring.

Derive the equations of motion and determine the equilibrium positions of the point mass relative to the ring.

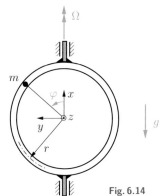

Fig. 6.14

Results: see (**B**) $-m(r\dot\varphi^2 \cos\varphi + r\ddot\varphi \sin\varphi) = -m\,g + N_x$,

$$-m(r\dot\varphi^2 \sin\varphi - r\ddot\varphi \cos\varphi + r\Omega^2 \sin\varphi) = N_x \tan\varphi,$$

$$2\,m\,r\,\Omega\,\dot\varphi \cos\varphi = N_z,\ \varphi_1 = 0,\ \varphi_2 = \pi,\ \varphi_{3,4} = \pi \pm \arccos\left(\frac{g}{r\Omega^2}\right).$$

Example 6.12 A point P moves on a square plate along a circular path (radius r) with a constant relative velocity v_r. The plate moves horizontally with the constant acceleration a_0 (Fig. 6.15).

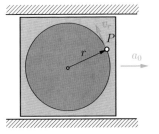

Determine the magnitude of the absolute acceleration of P.

Fig. 6.15

Result: see (**A**) $a = \sqrt{a_0^2 + \dfrac{v_r^4}{r^2} - 2a_0 \dfrac{v_r^2}{r} \cos \varphi}$

E6.12

Example 6.13 A crane starts to move from rest with the constant acceleration b_0 along straight tracks. At the same time, the extension arm begins to rotate with the constant angular velocity ω, and the crab on the extension arm begins to move towards point P with the constant relative acceleration b_c (Fig. 6.16). The initial positions of the extension arm and of the crab are given by φ_0 and s_0.

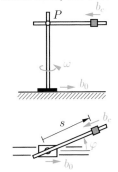

Determine the absolute velocity and the absolute acceleration of the crab as functions of the time t.

Fig. 6.16

E6.13

Results: see (**A**)

$$\boldsymbol{v} = [b_0 t \cos \varphi - b_c t]\boldsymbol{e}_\xi + [-b_0 t \sin \varphi + \omega(-\tfrac{1}{2}b_c t^2 + s_0)]\boldsymbol{e}_\eta\,,$$

$$\boldsymbol{a} = [b_0 \cos \varphi - \omega^2(-\tfrac{1}{2}b_c t^2 + s_0) - b_c]\boldsymbol{e}_\xi - [b_0 \sin \varphi + 2\,\omega\,b_c t]\boldsymbol{e}_\eta\,.$$

6.4 Summary

- Absolute velocity: $v_a = v_f + v_r$,

    $v_f = v_0 + \boldsymbol{\omega} \times r_{0P}$ fixed frame velocity,

    $v_r = \dfrac{\mathrm{d}^* r_{0P}}{\mathrm{d}t}$ relative velocity.

- Absolute acceleration: $a_a = a_f + a_r + a_c$,

    $a_f = a_0 + \dot{\boldsymbol{\omega}} \times r_{0P} + \boldsymbol{\omega} \times (\boldsymbol{\omega} \times r_{0P})$ fixed frame acceleration,

    $a_r = \dfrac{\mathrm{d}^* v_r}{\mathrm{d}t}$ relative acceleration,

    $a_c = 2\,\boldsymbol{\omega} \times v_r$ Coriolis acceleration.

- Newton's 2nd Law in a moving frame:

    $m\,a_r = F + F_f + F_c$,

    F force exerted on the point mass,

    $F_f = -m\,a_f$ D'Alembert's inertial force (fictitious force),

    $F_c = -m\,a_c$ Coriolis force (fictitious force).

- If the reference frame is translating with a constant velocity ($a_f = 0$, $a_c = 0$), then the equation of motion in the moving frame is identical with the equation of motion in a fixed frame.
- Fixed frames and uniformly moving frames are inertial frames.

Chapter 7

Numerical Simulation

7

7 Numerical Simulation

Objectives: So far we have discussed problems where the equations of motion could be solved analytically. In many applications it is difficult or even impossible to find an analytical solution. In these cases it is necessary to calculate an approximate solution with the help of an appropriate numerical integration scheme. We will discuss several numerical methods which enable us to compute numerical solutions of differential equations. You will learn how to apply numerical methods to treat initial-value problems.

7.1 Introduction

In the preceding chapters we have discussed the motion of point masses and rigid bodies in cases where we were able to solve the equations of motions analytically. In a variety of problems this is an elaborate undertaking or even impossible. An example is the equation of motion (5.19) of a simple pendulum, also called an ideal pendulum: $\ddot{\varphi} + (g/l)\sin\varphi = 0$. The analytical solution $\varphi(t)$ of this nonlinear differential equation for finite amplitudes involves the use of an elliptic integral, i.e., (5.19) cannot be solved with elementary functions. Only the usage of a small-angle approximation ($\varphi \ll 1$) leads to the much simpler form of a harmonic motion (cf. Section 5.2.1). Nevertheless, we are able to compute accurate approximate solutions for such complicated problems with the help of numerical methods.

In the following we analyse linear and non-linear differential equations of first and second order with assigned initial values. Thus we have what is called *initial-value problems of first* or *second order*. In this chapter we will introduce some basic numerical methods for solving ordinary differential equations for initial-value problems. A more detailed discussion of this topic is given in standard textbooks.

Most numerical solution methods are based on the computation of approximations for differential equations of first order. Therefore, we will start with numerical solution techniques for first-order systems. Higher-order differential equations can always be transformed into a system of first-order differential equations. This system can then be solved with the numerical techniques for first-order systems.

7.2 First-Order Initial-Value Problems

In many applications the equation of motion can be expressed by a differential equation of first order. As an example let us consider the acceleration a of a point mass as a function of the velocity v. Here, we have the general form $a = f(v)$ and we are interested in an approximate solution of the velocity $v(t)$ (cf. Section 1.1.3). In this case the equation of motion is formulated as $\dot{v} = f[v(t)]$ with the assigned initial value $v(t_s) = v_s$. Here t_s denotes the starting value of the time interval of interest and v_s is the associated initial velocity.

In the following we consider an ordinary differential equation of first order

$$\dot{x}(t) = f[t, x(t)] \tag{7.1}$$

over a time interval $[t_s, t_e]$ subjected to an *initial condition* $x_s = x(t_s)$. The basic idea of the numerical treatment of initial-value problems is the approximation of the time-dependent course of the functions $x(t)$ and $\dot{x}(t)$ at discrete points (mesh points). For this purpose we perform a time discretization and divide the considered time interval $[t_s, t_e]$ in n equally spaced time increments $\triangle t$ (Fig. 7.1). The *step size* is

$$\triangle t = \frac{t_e - t_s}{n} \, , \tag{7.2}$$

whereby a discrete point on the time axis is determined by

$$t_i = t_a + i \triangle t \quad \text{with} \quad i = 0, ..., n \, . \tag{7.3}$$

In order to simplify the notation we denote the approximate values at time t_i with

$$x_i = x(t_i) \quad \text{and} \quad \dot{x}_i = \dot{x}(t_i) \, . \tag{7.4}$$

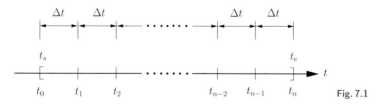

Fig. 7.1

The integration of (7.1) from the lower limit t_i to the upper limit t_{i+1} yields with

$$\int_{t_i}^{t_{i+1}} \dot{x}(t) \, \mathrm{d}t = x(t_{i+1}) - x(t_i)$$

the expression

$$x(t_{i+1}) = x(t_i) + \int_{t_i}^{t_{i+1}} f[t, x(t)] \, \mathrm{d}t . \tag{7.5}$$

Equation (7.5) is the basis for a variety of algorithms for the computation of approximate solutions x_{i+1}. These algorithms differ from each other by the choice for the approximation of the integral term.

The simplest method is *Euler's method* ("foward Euler"):

$$\int_{t_i}^{t_{i+1}} f[t, x(t)] \, \mathrm{d}t \approx f[t_i, x_i] \, \triangle t$$

$$\rightarrow \quad x_{i+1} = x_i + f[t_i, x_i] \, \triangle t . \tag{7.6}$$

If x_i is known, we are able to compute $f[t_i, x_i]$ by a simple evaluation of the given function and thus obtain x_{i+1}. This procedure is called *explicit* since Equation (7.6) yields immediately (explicitly) the unknown quantity x_{i+1}. An illustration of the procedure is depicted in Fig. 7.2a.

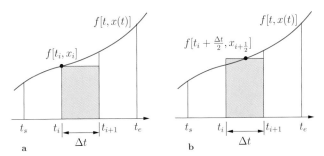

Fig. 7.2   a   b

The advantage of Euler's method is its simplicity of implementation. However, in order to achieve accurate approximations for practical problems often a small step size has to be used.

The recurrence formula (7.6) has to be evaluated for $i = 0, ..., n$. In order to illustrate the procedure we summarize the algorithm in Table 7.1 and set $t_s = 0$.

Table 7.1

i	t_i	x_i	$f[t_i, x_i]$
0	$t_0 = 0$	$x_0 = x_s$	$f[t_0, x_0]$
1	$t_1 = \triangle t$	$x_1 = x_0 + f[t_0, x_0]\triangle t$	$f[t_1, x_1]$
2	$t_2 = 2\triangle t$	$x_2 = x_1 + f[t_1, x_1]\triangle t$	$f[t_2, x_2]$
3	$t_3 = 3\triangle t$	$x_3 = x_2 + f[t_2, x_2]\triangle t$	$f[t_3, x_3]$
$\vdots$	$\vdots$	$\vdots$	$\vdots$
n	$t_n = n\triangle t$	$x_n = x_{n-1} + f[t_{n-1}, x_{n-1}]\triangle t$	$f[t_n, x_n]$

As can be seen, the process requires function evaluations row by row, which can easily be implemented.

A modification of Euler's method is based on the approximation of the integral (7.5) by means of the value of the function in the middle of the time interval $[t_i, t_{i+1}]$. In this case the approximate solution is given by

$$\int_{t_i}^{t_{i+1}} f[t, x(t)]\, \mathrm{d}t \approx f[t_i + \frac{\triangle t}{2}, x_{i+\frac{1}{2}}]\,\triangle t$$

$$\rightarrow \quad x_{i+1} = x_i + f[t_i + \frac{\triangle t}{2}, x_{i+\frac{1}{2}}]\,\triangle t\,. \quad (7.7)$$

The value $x_{i+\frac{1}{2}}$ is determined by the explicit computation

$$x_{i+\frac{1}{2}} = x_i + f[t_i, x_i]\,\frac{\triangle t}{2}\,. \tag{7.8}$$

With the abbreviations

$$k_1 = f[t_i, x_i] \quad \text{and} \quad k_2 = f[t_i + \frac{\triangle t}{2}, x_i + k_1 \frac{\triangle t}{2}] \tag{7.9}$$

we obtain the approximate value at t_{i+1} using the formula

$$x_{i+1} = x_i + k_2\,\triangle t\,. \tag{7.10}$$

This procedure is known as the second-order *Runge-Kutta method* (Carle David Tolmé Runge 1856-1927, Martin Wilhelm Kutta 1867-1944), since the error is of the order of $\triangle t^2$. The procedure is also known as the *midpoint method*; for a visualization see Fig. 7.2b.

Another algorithm is the fourth-order Runge-Kutta method. Herein four function evaluations are used for the computation of an approximate solution. As in the aforementioned midpoint method we consider results of several explicit computations performed with Euler's method. The individual steps within the fourth-order Runge-Kutta procedure may be summarized as follows:

$$x_{i+1} = x_i + k \, \triangle t \quad \text{with} \quad k = \frac{1}{6}(k_1 + 2k_2 + 2k_3 + k_4) \quad (7.11)$$

where we have used the abbreviations

$$k_1 = f[t_i, x_i] \, ,$$

$$k_2 = f[t_i + \frac{\triangle t}{2}, x_i + \frac{\triangle t}{2} k_1] \, ,$$

$$k_3 = f[t_i + \frac{\triangle t}{2}, x_i + \frac{\triangle t}{2} k_2] \, ,$$

$$k_4 = f[t_i + \triangle t, x_i + \triangle t \, k_3] \, .$$

Table 7.2

i	t_i	x_i	k_1	k_2	k_3	k_4	k
0	t_0	x_0	$f[t_0, x_0]$	$f[\bar{t}_0, x_0 + \frac{\triangle t}{2} k_1]$	$f[\bar{t}_0, x_0 + \frac{\triangle t}{2} k_2]$	$f[t_0 + \triangle t, x_0 + \triangle t \, k_3]$	k
1	t_1	x_1	$f[t_1, x_1]$	$f[\bar{t}_1, x_1 + \frac{\triangle t}{2} k_1]$	$f[\bar{t}_1, x_1 + \frac{\triangle t}{2} k_2]$	$f[t_1 + \triangle t, x_1 + \triangle t \, k_3]$	k
$\vdots$	$\vdots$	$\vdots$	$\vdots$	$\vdots$	$\vdots$	$\vdots$	$\vdots$
n	t_n	x_n	$f[t_n, x_n]$	$f[\bar{t}_n, x_n + \frac{\triangle t}{2} k_1]$	$f[\bar{t}_n, x_n + \frac{\triangle t}{2} k_2]$	$f[t_n + \triangle t, x_n + \triangle t \, k_3]$	k

The sequence of the individual computational steps is summarized in Table 7.2. First we set the starting values for time t and the variable x, i.e., we initialize $t_0 = t_s$ and set the initial value $x_0 = x_s$. Subsequently we compute the values k_1, k_2, k_3, k_4 and k (Table 7.2, first row). Now we can compute the approximation x_1 as depicted in the second row of the table by evaluating (7.11). This procedure has to be repeated until the end of the time interval t_e is reached. In Table 7.2 we have introduced the abbreviation $\tilde{t}_i = t_i + \triangle t/2$.

Euler's method as well as Runge-Kutta's method can easily be applied to systems of differential equations (Appendix A).

As a first example for the application of Euler's method we consider the translational motion of a car (mass $m = 1400$ kg), see Fig. 7.3. The constant driving force is $F = 2800$ N and the aerodynamic drag is assumed to be $F_d = c_d v^2$ (drag coefficient $c_d = 0.7$). We are interested in an approximate solution for the velocity in the time interval $[t_s, t_e] = [0, 120 \text{ s}]$; the initial velocity is set to $v_s = 10$ m/s.

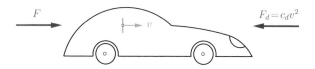

F $\qquad$ v $\qquad$ $F_d = c_d v^2$

Fig. 7.3

The equation of motion is

$$m\dot{v} = F - c_d v^2 \quad \rightarrow \quad \dot{v} = f[t, v] \quad \text{with} \quad f[t, v] = \frac{F}{m} - \frac{c_d}{m} v^2 .$$

Using $\kappa^2 = F/c_d$ the analytical solution is given by

$$\dot{x} = \kappa \tanh\frac{F(t - C)}{m\kappa} \quad \text{with} \quad C = -\frac{m\kappa}{F} \operatorname{artanh}\frac{v_s}{\kappa} ,$$

see Section 1.2.4. At time $t = 40$ s the exact analytical value for the velocity is $\dot{x}(40) = 56.32 \approx 56.3$ m/s, at time t_e we get the exact value $\dot{x}(120) = 63.199 \approx 63.2$ m/s.

In the following numerical simulations we will consistently use the units N, m and s. As is common in computerbased calculations, we will skip the units for intermediate solutions. We always have to write down all quantities in the units given above before we can run a simulation.

With the given parameters we obtain

$$\dot{v} = f[t, v] \quad \text{with} \quad f[t, v] = 2 - 0.0005 \, v^2 .$$

In order to explain the recurrence procedure in detail, we perform the calculation with 6 equidistant time steps:

$$t_i = i \, \triangle t \quad \text{with} \quad i = 0, ..., 6 \quad \rightarrow \quad \triangle t = \frac{120 - 0}{6} = 20 .$$

For $t = 0$, i.e., for $i = 0$ we obtain the values

$$v_0 = v_a = 10 ,$$

$$f[t_0, v_0] = 2 - 0.0005 \cdot 10^2 = 1.950 .$$

At time $t_1 = 20$ $(i = 1)$ we get

$$v_1 = v_0 + f[t_0, v_0] \, \triangle t = 10 + 1.95 \cdot 20 = 49 ,$$

$$f[t_1, v_1] = 2 - 0.0005 \cdot 49^2 = 0.7995 .$$

The evaluation at time $t_2 = 40$ $(i = 2)$ leads to

$$v_2 = v_1 + f[t_1, v_1] \, \triangle t = 49 + 0.7995 \cdot 20 = 64.99 ,$$

$$f[t_2, v_2] = 2 - 0.0005 \cdot 64.99^2 = -0.1119 .$$

In this manner we compute the values for $i = 4, ..., 6$. The results are summarized in Table 7.3.

Table 7.3

i	t_i	v_i	$f[t_i, v_i]$
0	0	10.000	1.9500
1	20	49.000	0.7995
2	40	64.990	-0.1119
3	60	62.753	0.0310
4	80	63.374	-0.0081
5	100	63.211	0.0022
6	120	63.255	-0.0006

A comparison of the approximate solution with the analytical solution at time $t = 40$ s shows a deviation of about 15%, whereas at time $t = 120$ s the error is smaller than 1%, see Fig. 7.4.

The implementation is shown in Algorithm 7.1.

Algorithm 7.1: Euler's Method

```
% parameter
cd = 0.7;   m = 1400.0;   F = 2800.0;
% number of time steps, time interval, time increment
n = 6; ts = 0.0; te = 120.0;
delta_t = (te - ts)/n;
% initial condition
v(1) = 10.0;    t(1) = ts;
for i = 1 : n+1
   func = F/m - cd/m*(v(i))^2;
   v(i+1) = v(i) + func*delta_t;
   t(i+1) = t(i) + delta_t;
end
```

Fig. 7.4 depicts the numerical solutions for $n = 6, 10, 20$ time increments. It can be seen that for a larger number of time increments the solution converges to the exact solution. For $n = 20$ the solution is more or less identical with the exact solution.

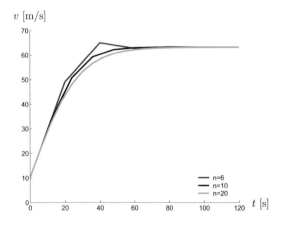

Fig. 7.4

E7.1

Example 7.1 Compute an approximate solution for the velocity $v(t)$ of the preceding example with the fourth-order Runge-Kutta method.

Solution We divide the time interval into 6 equidistant increments:

$$t_i = i \triangle t \quad \text{with} \quad i = 0, ..., 6 \quad \rightarrow \quad \triangle t = \frac{120 - 0}{6} = 20 \ .$$

At time $t = 0$ $(i = 0)$ we set the initial condition $v_0 = v_s = 10$. Subsequently we compute the values k_1, k_2, k_3, k_4:

$$k_1 = f[t_0, v_0] = 2 - 0.0005 \cdot 10^2 = 1.950 \ ,$$

$$k_2 = f[t_0 + \frac{\triangle t}{2}, v_0 + \frac{\triangle t}{2}k_1]$$
$$= 2 - 0.0005 \cdot (10 + 20/2 \cdot 1.950)^2 = 1.5649 \ ,$$

$$k_3 = f[t_0 + \frac{\triangle t}{2}, x_0 + \frac{\triangle t}{2}k_2]$$
$$= 2 - 0.0005 \cdot (10 + 20/2 \cdot 1.5649)^2 = 1.6711 \ ,$$

$$k_4 = f[t_i + \triangle t, x_i + \triangle t \, k_3]$$
$$= 2 - 0.0005 \cdot (10 + 20 \cdot 1.6711)^2 = 1.0573 \ .$$

Thus the k-value is

$$k = \frac{1}{6}(k_1 + 2k_2 + 2k_3 + k_4) = 1.5799 \ ,$$

and we obtain at $t = 20$ $(i = 1)$ the velocity

$$v_1 = v_0 + k \triangle t \quad \rightarrow \quad v_1 = 10 + 1.5799 \cdot 20 = 41.597 \ .$$

This procedure has to be repeated until the end of the time interval is reached. We summarize the computed values for a discretization of the time interval $[t_s, t_e]$ with $n = 6$ equidistant time increments in Table 7.4.

A comparison of the approximate solution with the exact one at the discrete times $t = 40$ s and $t = 120$ s reveals that the error is less than 1% in each case.

The implementation is depicted in Algorithm 7.2.

Table 7.4

i	t_i	v_i	k_1	k_2	k_3	k_4	k
0	0	10.000	1.9500	1.5649	1.6711	1.0573	1.5799
1	20	41.597	1.1348	0.5989	0.8680	0.2620	0.7216
2	40	56.029	0.4304	0.1800	0.3279	0.0414	0.2479
3	60	60.988	0.1403	0.0537	0.1073	0.0070	0.0782
4	80	62.552	0.0436	0.0162	0.0334	0.0016	0.0241
5	100	63.034	0.0134	0.0049	0.0103	0.0004	0.0074
6	120	63.181	0.0041	0.0015	0.0031	0.0001	0.0022

Algorithm 7.2: Fourth-order Runge-Kutta method

```
% initialization of
% parameters and functions in subroutine func.m
%
% number of equidistant time increments
n = 6;
% time interval
ts = 0.0; te = 120.0;
% step size, i.e., time increment
delta_t = (te - ts)/n;
% initial condition
v(1) = 10.0, t(1) = ts,
% recurrence formula
for i = 1 : n+1
  k1 = func(v(i));
  k2 = func(v(i) + delta_t/2*k1);
  k3 = func(v(i) + delta_t/2*k2);
  k4 = func(v(i) + delta_t  *k3);
  k  = 1/6*(k1 + 2*k2 + 2*k3 + k4);
  v(i+1) = v(i) + k*delta_t;
  t(i+1) = t(i) + delta_t;
end
```

```
% initialization of parameters and functions in func.m
function value = func(v)
cd = 0.7;   m = 1400.0;  F = 2800.0;
value = F/m - cd/m*v^2;
```

The approximate solutions which have been computed with the fourth-order Runge-Kutta method for $n = 6, 10$ and 20 are shown in Fig. 7.5. In contrast to Euler's method we already obtain accurate solutions for $n = 6$ increments.

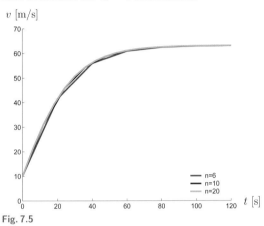

Fig. 7.5

7.3 Second-Order Initial-Value Problems

In kinetics the principle of linear momentum $m\,\ddot{x} = F$ and the principle of angular momentum $\Theta\,\ddot{\varphi} = M$ lead to equations of motion of the form $a\,\ddot{x} = b$. In general the equation of motion and the associated *initial conditions* of position and velocity are

$$\ddot{x} = f[t, x(t), \dot{x}(t)] \quad \text{with} \quad x_s = x(t_s) \quad \text{and} \quad \dot{x}_s = \dot{x}(t_s) . \quad (7.12)$$

In order to apply the discretization procedures for first-order initial-value problems (see Section 7.2) to (7.12) we have to transform the equation of motion (7.12) to a system of two differential equations of first order (cf. Appendix A). Therefore, we introduce the functions $z_1(t)$ and $z_2(t)$. Applying the transformations

$$z_1(t) = x(t) \quad \text{and} \quad z_2(t) = \dot{x}(t)$$

yields the system of differential equations

$$\dot{z}_1(t) = f_1[t, z_1(t), z_2(t)] \ ,$$

$$\dot{z}_2(t) = f_2[t, z_1(t), z_2(t)] \ ,$$

$$(7.13)$$

where $f_1[t, z_1(t), z_2(t)] = z_2(t)$. These two differential equations can now be solved by means of the discretization procedures presented in Section 7.2. For reasons of clarity and comprehensibility we introduce the following abbreviations for the approximate values $z_1(t)$ and $z_2(t)$ at the discrete times t_i:

$$z_{1_i} := z_1(t_i) \quad \text{and} \quad z_{2_i} := z_2(t_i) \quad \text{for} \quad i = 0, ..., n \ . \quad (7.14)$$

The initial conditions are

$$z_{1_0} = z_1(t_s) = x_s \quad \text{and} \quad z_{2_0} = z_2(t_s) = \dot{x}_s \ . \quad (7.15)$$

The recurrence formula for *Euler's method* (cf. (7.6)) is

$$z_{1_{i+1}} = z_{1_i} + f_1[t_i, z_{1_i}, z_{2_i}] \, \triangle t \ ,$$

$$z_{2_{i+1}} = z_{2_i} + f_2[t_i, z_{1_i}, z_{2_i}] \, \triangle t \ .$$

$$(7.16)$$

Applying the *fourth-order Runge-Kutta method* to the integration of first-order initial-value problems requires the evaluation of the four values k_1, k_2, k_3, k_4. In a second-order problem we obtain two differential equations of first order, therefore we also have to compute four further values for the second equation. They are denoted by l_1, l_2, l_3, l_4. Using the abbreviation $\tilde{t}_i = t_i + \triangle t/2$ we have the following expressions for the function evaluations k_j, l_j for $j = 1, ..., 4$:

$$k_1 = f_1[t_i, z_{1_i}, z_{2_i}],$$

$$l_1 = f_2[t_i, z_{1_i}, z_{2_i}] \ ,$$

$$k_2 = f_1[\tilde{t}_i, z_{1_i} + \frac{\triangle t}{2} k_1, z_{2_i} + \frac{\triangle t}{2} l_1],$$

$$l_2 = f_2[\tilde{t}_i, z_{1_i} + \frac{\triangle t}{2} k_1, z_{2_i} + \frac{\triangle t}{2} l_1] \ ,$$

$$k_3 = f_1[\tilde{t}_i, z_{1_i} + \frac{\triangle t}{2}\, k_2, z_{2_i} + \frac{\triangle t}{2}\, l_2],$$

$$l_3 = f_2[\tilde{t}_i, z_{1_i} + \frac{\triangle t}{2}\, k_2, z_{2_i} + \frac{\triangle t}{2}\, l_2]\ ,$$

$$k_4 = f_1[t_{i+1}, z_{1_i} + \triangle t\, k_3, z_{2_i} + \triangle t\, l_3],$$

$$l_4 = f_2[t_{i+1}, z_{1_i} + \triangle t\, k_3, z_{2_i} + \triangle t\, l_3]\ .$$

Finally we have to compute the parameters

$$k = \frac{1}{6}(k_1 + 2k_2 + 2k_3 + k_4) \quad \text{and} \quad l = \frac{1}{6}(l_1 + 2l_2 + 2l_3 + l_4)$$

$$(7.17)$$

for the calculation of the approximate solutions at time t_{i+1}. The individual calculation steps are summarized in Table 7.5.

Table 7.5

$z_{1_{i+1}} = z_{1_i} + k\, \triangle t$	$z_{2_{i+1}} = z_{2_i} + l\, \triangle t$
$k = \frac{1}{6}(k_1 + 2k_2 + 2k_3 + k_4)$	$l = \frac{1}{6}(l_1 + 2l_2 + 2l_3 + l_4)$
$k_1 = f_1(t_i, z_{1_i}, z_{2_i})$	$l_1 = f_2(t_i, z_{1_i}, z_{2_i})$
$k_2 = f_1(\tilde{t}_i, z_{1_i} + \frac{\triangle t}{2}\, k_1, z_{2_i} + \frac{\triangle t}{2}\, l_1)$	$l_2 = f_2(\tilde{t}_i, z_{1_i} + \frac{\triangle t}{2}\, k_1, z_{2_i} + \frac{\triangle t}{2}\, l_1)$
$k_3 = f_1(\tilde{t}_i, z_{1_i} + \frac{\triangle t}{2}\, k_2, z_{2_i} + \frac{\triangle t}{2}\, l_2)$	$l_3 = f_2(\tilde{t}_i, z_{1_i} + \frac{\triangle t}{2}\, k_2, z_{2_i} + \frac{\triangle t}{2}\, l_2)$
$k_4 = f_1(t_{i+1}, z_{1_i} + \triangle t\, k_3, z_{2_i} + \triangle t\, l_3)$	$l_4 = f_2(t_{i+1}, z_{1_i} + \triangle t\, k_3, z_{2_i} + \triangle t\, l_3)$

A comparison of both columns in Table 7.5 shows that the columns can be interchanged when we substitute $k_i \leftrightarrow l_i$ as well as $f_1 \leftrightarrow f_2$. Therefore we rename

$$\{k_1, k_2, k_3, k_4\} \quad \rightarrow \quad \{k_{11}, k_{12}, k_{13}, k_{14}\}$$

$$\{l_1,\ l_2,\ l_3,\ l_4\} \quad \rightarrow \quad \{k_{21}, k_{22}, k_{23}, k_{24}\}\ .$$

Now the algorithm is given with $j = 1, 2$ as shown in Table 7.6.

Table 7.6

$$z_{j_{i+1}} = z_{j_i} + k_j \, \triangle t$$

$$k_j \quad = \tfrac{1}{6}(k_{j1} + 2k_{j2} + 2k_{j3} + k_{j4})$$

$$k_{j1} \quad = f_j(t_i, \ z_{1_i}, \ z_{2_i})$$

$$k_{j2} \quad = f_j(\tilde{t}_i, \ z_{1_i} + \tfrac{\triangle t}{2} k_{11}, \ z_{2_i} + \tfrac{\triangle t}{2} k_{21})$$

$$k_{j3} \quad = f_j(\tilde{t}_i, \ z_{1_i} + \tfrac{\triangle t}{2} k_{12}, \ z_{2_i} + \tfrac{\triangle t}{2} k_{22})$$

$$k_{j4} \quad = f_j(t_{i+1}, \ z_{1_i} + \triangle t \ k_{13}, \ z_{2_i} + \triangle t \ k_{23})$$

The implementation of this procedure is presented in Algorithm 7.3.

Algorithm 7.3: Fourth-order Runge-Kutta method

```
% fourth-order Runge-Kutta method
% for initial-value problems of second order
%
% initialization of
% parameters and functions in func1.m and func2.m:

% initialization of func1.m
function value1 = f1(t,z1,z2)
value1 = z2;

% initialization of func2.m
function value2 = f2(t,z1,z2)
value2 = ***;

% number of equidistant time increments
n = ***;
% time interval
ts = ***;
te = ***;
```

```
% step size, i.e., time increment
delta_t = (te - ts)/n;
% initial conditions
t(1) = ts;
z1(1) = ***;
z2(1) = ***;

% recurrence formula
for i = 1 : n+1
    k(1,1) = func1(t(i),z1(i),z2(i));
    k(2,1) = func2(t(i),z1(i),z2(i));
    k(1,2) = func1(t(i)+delta_t/2,
             z1(i)+delta_t/2*k(1,1),z2(i)+delta_t/2*k(2,1));
    k(2,2) = func2(t(i)+delta_t/2,
             z1(i)+delta_t/2*k(1,1),z2(i)+delta_t/2*k(2,1));
    k(1,3) = func1(t(i)+delta_t/2,
             z1(i)+delta_t/2*k(1,2),z2(i)+delta_t/2*k(2,2));
    k(2,3) = func2(t(i)+delta_t/2,
             z1(i)+delta_t/2*k(1,2),z2(i)+delta_t/2*k(2,2));
    k(1,4) = func1(t(i)+delta_t,
             z1(i)+delta_t*k(1,3),z2(i)+delta_t*k(2,3));
    k(2,4) = func2(t(i)+delta_t,
             z1(i)+delta_t*k(1,3),z2(i)+delta_t*k(2,3));
    z1(i+1) = z1(i)+1/6*(k(1,1)+2*k(1,2)
            + 2*k(1,3)+k(1,4))*delta_t;
    z2(i+1) = z2(i)+1/6*(k(2,1)+2*k(2,2)
            + 2*k(2,3)+k(2,4))*delta_t;
    t(i+1) = t(i)+delta_t;
end
```

For the numerical solution of different initial-value problems only the marked quantities *** have to be modified.

In a sequence of four illustrative examples we will solve the equation of motion

$$m\ddot{x} + d\dot{x} + kx = F_0 \cos \Omega t$$

of an oscillator with one degree of freedom, see Fig. 7.6, for different parameters (m, d, c, F_0, Ω) as well as for different initial conditions $x_s = x(t_s)$ and $\dot{x}_s = \dot{x}(t_s)$. To compute the approximate solutions for the position $x(t)$ and the velocity $v(t)$ in the time interval $[t_s, t_e]$ we use the fourth-order Runge-Kutta method.

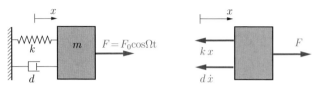

Fig. 7.6

First we write the equation of motion in the form

$$\ddot{x} = \frac{F_0}{m} \cos \Omega t - \frac{d}{m} \dot{x} - \frac{k}{m} x \ .$$

Introducing the functions

$$z_1(t) = x(t) \quad \text{and} \quad z_2(t) = \dot{x}(t)$$

we obtain two differential equations of first order

$$\dot{z}_1(t) = z_2(t) \ , \qquad \dot{z}_2(t) = \frac{F_0}{m} \cos \Omega t - \frac{d}{m} z_2(t) - \frac{k}{m} z_1(t) \ .$$

The functions $f_1[t, z_1(t), z_2(t)]$ and $f_2[t, z_1(t), z_2(t)]$, as introduced in (7.13), are

$$f_1[t, z_1(t), z_2(t)] = z_2(t) \ ,$$
$$f_2[t, z_1(t), z_2(t)] = \frac{F_0}{m} \cos \Omega t - \frac{d}{m} z_2(t) - \frac{k}{m} z_1(t) \ .$$

The initial conditions at time t_s are denoted as

$$z_1(t_s) = x_s \quad \text{and} \quad z_2(t_s) = \dot{x}_s \ .$$

In the first example we analyse an undamped free vibration $(d = 0, F_0 = 0)$ with the initial conditions

$$x_s = 0.1 \text{ m} \quad \text{and} \quad \dot{x}_s = 0 \ .$$

The parameters are set to

$$m = 5 \text{ kg} \quad \text{and} \quad k = 500 \text{ N/m} \ .$$

The analysis is performed using $n = 20$ and $n = 40$ time increments. The modifications in Algorithm 7.3 are:

```
% number of equidistant time increments
n = 20;
% time interval, time increment
ts = 0.0;  te = 3.0;  delta_t = (te - ts)/n;
% initial conditions
z1(1) = 0.1;  z2(1) = 0.0;  t(1) = ts;


% initialization of func1.m
function value1 = f1(t,z1,z2)
value1 = z2;
% initialization of func2.m
function value2 = f2(t,z1,z2)
m = 5.0; kf = 500.0;
value2 = -kf/m*z1;
```

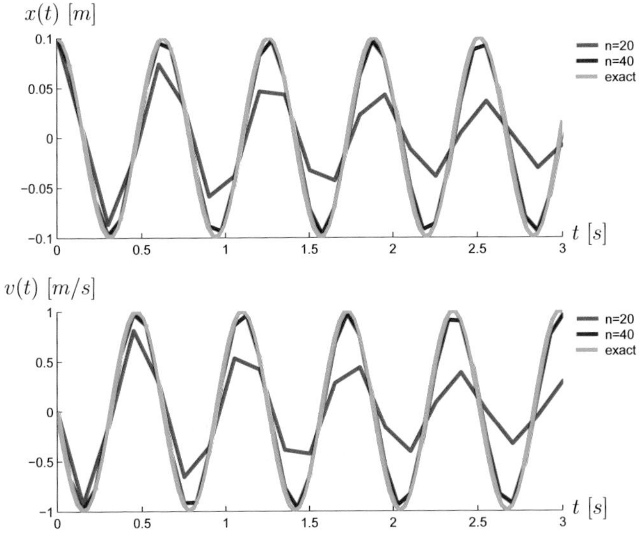

Fig. 7.7

The numerical results for the position-time diagram and the velocity-time diagram are shown in Fig. 7.7. For $n = 40$ we obtain a good agreement with the analytical solution $x = 0.1 \cos \omega t$ where $\omega = \sqrt{k/m} = \sqrt{500/5} = 10 \text{ s}^{-1}$ and $T = 2\pi/\omega = 0.63$ s.

In the second example we consider a damped free vibration with the parameters

$$m = 5 \text{ kg}, \quad k = 500 \text{ N/m}, \quad d = 10 \text{ kg/s}$$

and the initial conditions $x_s = 0.1$ m, $\dot{x}_s = 0$. The necessary modifications in Algorithm 7.3 are:

```
% number of equidistant time increments
n = 20;
% time interval, time increment
ts = 0.0;   te = 3.0;   delta_t = (te - ts)/n;
% initial conditions
z1(1) = 0.1;   z2(1) = 0.0;   t(1) = ts;

% initialization of func1.m
function value1 = f1(t,z1,z2)
value1 = z2;
% initialization of func2.m
function value2 = f2(t,z1,z2)
m = 5.0; d = 10.0; kf = 500.0;
value2 = -d/m*z2-kf/m*z1;
```

The numerical solutions for $n = 20$ and $n = 40$ are depicted in Fig. 7.8 and compared with the analytical solution. We obtain a good agreement between the numerical simulation calculated with $n = 40$ time increments and the analytical one.

In the third example we consider an aperiodic motion of a critically damped system. In contrast to the situation above we introduce a larger damping coefficient d and set $t_e = 1$ s. The critical damping coefficient is obtained from $\zeta = \xi/\omega = 1 \rightarrow \xi = \omega = 10 \text{ s}^{-1}$ and has the value $d = 2\xi\, m = 100$ kg/s. All other settings remain the same as in the previous example.

Fig. 7.9 shows the results of the numerical simulation. Obviously, we obtain accurate approximate solutions for $n = 20$ as well as for $n = 40$.

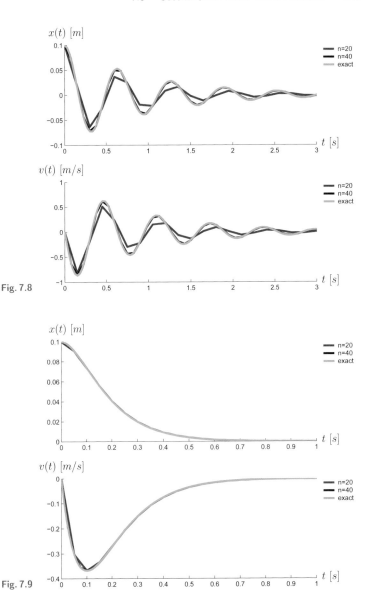

Fig. 7.8

Fig. 7.9

In the fourth example we analyse the vibration of a damped system subjected to a harmonic force and the initial conditions

$$x_s = 0 \quad \text{and} \quad \dot{x}_s = 0 .$$

The parameters are chosen as

$$d = 10 \text{ kg/s}, \quad k = 500 \text{ N/m}, \quad F_0 = 10 \text{ N}, \quad \Omega = 10 \text{ s}^{-1}$$

and the simulation is carried out for $n = 30$ and $n = 50$ increments. Algorithm 7.3 is therefore modified as follows:

```
% number of equidistant time increments
n = 50;
% time interval, time increment
ts = 0.0;   te = 6.0;   delta_t = (te - ts)/n;
% initial conditions
z1(1) = 0.0;   z2(1) = 0.0;   t(1) = ta;

% initialization of func1.m
function value1 = f1(t,z1,z2)
value1 = z2;
% initialization of func2.m
function value2 = f2(t,z1,z2)
m = 5.0; d = 10.0; kf = 500.0; F_o = 10.0; Omega = 10.0;
value2 = -d/m*z2-kf/m*z1;
```

The numerical solutions for $n = 30$ and $n = 50$ are shown in Fig. 7.10 and compared with the analytical solution. As expected in the steady-state, the (response) vibration of the system has the frequency of the excitation Ω, i.e., it has the period of vibration $T = 2\pi/\Omega \approx 0.63$ s.

In the simulation the steady-state is reached after approximately 4 s. Then we can determine the amplitude using the magnification factor (5.62) and setting the parameter $E = 1$. For an excitation in resonance, i.e., $\eta = \Omega/\omega = 1$, the magnification factor takes the value $V_1(1) = 1/2\zeta \rightarrow V_1(1) = 1/0.2 = 5$. The maximum deflection is obtained from $x_{\max} = \pm V_1 x_0$ and takes with $x_0 = F_0/k = 10/500 = 0.02$ the value $x_{\max} = \pm 0.1$ m.

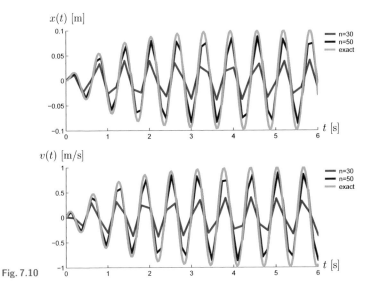

Fig. 7.10

Example 7.2   A pendulum supported at point A consists of a rod (length $2l$, with negligible mass) and two point masses m_1 and m_2, see Fig. 7.11. The pendulum undergoes a free vibration. Determine the angle-time diagrams $\varphi(t)$ in the time interval $[0,\ 43\,\text{s}]$ for the initial conditions $\varphi(0) = 3°$, $\varphi(0) = 179°$, $\varphi(0) = 179.99°$ all with $\dot{\varphi}(0) = 0$. Also plot the oscillations over one period of vibration T.

E7.2

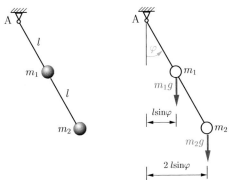

Fig. 7.11

Solution The equation of motion was derived in Example 2.3 using the principle of angular momentum:

$$\ddot{\varphi} + \omega^2 \sin\varphi = 0 \qquad \text{with} \qquad \omega^2 = \frac{g}{l}\frac{m_1 + 2m_2}{m_1 + 4m_2}.$$

The initial conditions are

$$\varphi_s = \varphi(0) \quad \text{and} \quad \dot{\varphi}_s = \dot{\varphi}(0) = 0.$$

In the numerical simulation we set $\omega = 1$. The required modifications in the Matlab-code of Algorihm 7.3 are:

```
% number of equidistant time increments
n = 3000;
% time interval and time increment
ts = 0.0;  te = 42.935;  delta_t = (te - ts)/n;
% initial conditions
z1(1) = (179.99/180)*pi;  z2(1) = 0.0;  t(1) = ts;

% initialization of func1.m
function value = func1(t,z1,z2)
value = z2;

% initialization of func2.m
function value = func2(t,z1,z2)
omega = 1;
value = -omega*omega*sin(z1);
```

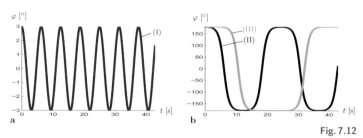

Fig. 7.12

The oscillation for the initial condition $\varphi(0) = 3°$ is depicted in Fig. 7.12a (I); we read off the diagram the period of vibration $T \approx 6.5$ s. Fig. 7.12b shows the diagrams for the initial angles

$\varphi(0) = 179°$ with $T \approx 24.5$ s (II) and $\varphi(0) = 179.99°$ with $T \approx 43$ s (III).

The oscillation corresponding to the inital condition $\varphi(0) = 179°$ exhibits the formation of a plateau in the domain of the maximal amplitude. This effect is more pronounced in the case of the initial condition $\varphi(0) = 179.99°$, see Fig. 7.12b. For large initial angles the function $\varphi(t)$ is the so-called *sinus amplitudinis* (Jacobi elliptic function, a generalization of the sine function), which differs strongly from the classical sine function.

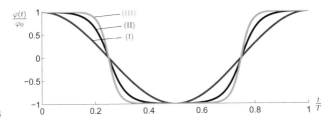

Fig. 7.13

These deviations can be observed much better, when we plot normalized $\varphi(t)$-diagrams. Therefore, we plot as the ordinate the quantity $\varphi(t)/\varphi_0$ and the abscissae is assigned to the actual time divided by the period of vibration (t/T). Fig. 7.13 shows the normalized angle-time diagrams.

The periods of vibration can also be determined analytically. In order to do this we multiply the equation of motion by $\dot{\varphi}$ and integrate with respect to time:

$$\int_t \ddot{\varphi}\dot{\varphi}\,\mathrm{d}t + \int_t \omega^2 \sin\varphi\dot{\varphi}\,\mathrm{d}t = C$$

$$\rightarrow \quad \frac{1}{2}\dot{\varphi}^2 + \int_0^\varphi \omega^2 \sin\varphi\,\mathrm{d}\varphi = C\,.$$

The integration over φ yields

$$\frac{1}{2}\dot{\varphi}^2 + \omega^2(1 - \cos\varphi) = C\,. \tag{a}$$

At φ_a we have $\dot{\varphi}(\varphi_a) = 0$; thus we obtain

$$C = \omega^2(1 - \cos\varphi_a)\,.$$

Substituting this result into (a) yields the angular velocity

$$\dot{\varphi} = \omega\sqrt{2(\cos\varphi - \cos\varphi_s)}\,.$$

Referring to (1.18) the period of vibration T is given by

$$T = \frac{4}{\omega}\int_0^{\varphi_0}\frac{\mathrm{d}\varphi}{\dot{\varphi}} = \frac{4}{\omega}\int_0^{\varphi_0}\frac{\mathrm{d}\varphi}{\sqrt{2(\cos\varphi - \cos\varphi_0)}}\,.$$

The numerical values for the periods of vibration, corresponding to the individual initial conditions, are

$$\varphi(0) = 3° : \qquad T = 6.284 \text{ s},$$
$$\varphi(0) = 179° : \qquad T = 24.511 \text{ s},$$
$$\varphi(0) = 179.99° : T = 42.935 \text{ s}.$$

7.4 Supplementary Examples

Detailed solutions to the following examples are given in (**C**) D. Gross et al. *Formeln und Aufgaben zur Technischen Mechanik 4*, Springer, Berlin 2008.

Example 7.3 A point mass m is thrown from a height $h = 5$ m with an initial velocity $v_0 = 10$ m/s at an angle $\alpha = 45°$ with respect to the horizontal. The only force acting on the mass is its weight $W = mg$ ($g \approx 10$ m/s^2).

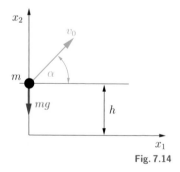

Fig. 7.14

Determine the trajectory $x_1(t)$, $x_2(t)$ within the time interval $[t_s; t_e] = [0; 2\text{s}]$ using Euler's method (number of time increments $n = 5, 10$ and 50). Program a short algorithm for the velocity and position updates and compare the results with the analytical solution.

Results: see (**C**)   Equation of motion:

$$\boldsymbol{F} = m\,\boldsymbol{a} \quad \text{with} \quad \boldsymbol{F} = \begin{pmatrix} 0 \\ -mg \end{pmatrix}, \quad \boldsymbol{a} = \begin{pmatrix} a_1 \\ a_2 \end{pmatrix} = \begin{pmatrix} 0 \\ -g \end{pmatrix}.$$

Initial velocities and initial positions:

$$\boldsymbol{v}_0 = \begin{pmatrix} v_1(0) \\ v_2(0) \end{pmatrix} = \begin{pmatrix} v_0 \ \cos\alpha \\ v_0 \ \sin\alpha \end{pmatrix}, \quad \boldsymbol{x}_0 = \begin{pmatrix} x_1(0) \\ x_2(0) \end{pmatrix} = \begin{pmatrix} 0 \\ h \end{pmatrix}.$$

Analytical solution: $\boldsymbol{x}(t) = \left(v_0\, t\, \cos\alpha,\ -\tfrac{1}{2}gt^2 + v_0\, t\, \sin\alpha + h\right)^T$.

Outline of the algorithmic treatment:

```
% parameters
ts = 0.0; te = 2.0; n = 5; alpha = pi/4; h = 5; v0 = 10;
% components of the acceleration vector
a1 = 0; a2 = -10.0;
% time increment
delta_t = (te - ts)/n;
% initial conditions for position and velocity
x1(1) = 0; x2(1) = h;
v1(1) = v0*cos(alpha); v2(1) = v0*sin(alpha);
% integration scheme
for k = 1:n
    x1(k + 1) = x1(k) + v1(k)*delta_t;
    x2(k + 1) = x2(k) + v2(k)*delta_t;
    v1(k + 1) = v1(k) + a1*delta_t;
    v2(k + 1) = v2(k) + a2*delta_t;
end
```

Approximate solutions:

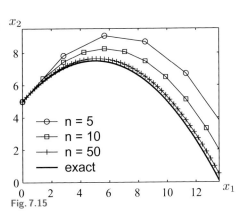

Fig. 7.15

Example 7.4 A rod of length l (with negligible mass) carries a point mass at its end.

Case i): Determine the $\varphi(t)$-diagram for $|\varphi| \ll 1$. Initial conditions: $\varphi(t_0) = 0.1$ and $\dot{\varphi}(t_0) = 0$.

Case ii): Compute the phase diagram ($\dot{\varphi}, \varphi$-diagram) for large deflections. Initial conditions: $\varphi(t_0) = \pi/2$ and $\dot{\varphi}(t_0) = 0$.

Use Euler's method for the numerical simulations.

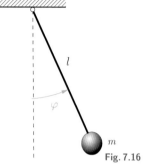

Fig. 7.16

Results: see (**C**)

Case i): Equation of motion (second-order differential equation):

$$\ddot{\varphi} + \frac{g}{l}\,\varphi = 0 \ .$$

Transformation with $z_1(t) = \varphi(t)$ and $z_2(t) = \dot{\varphi}(t)$ to two differential equations of first-order:

$$\dot{z}_1(t) = z_2(t) \quad \text{and} \quad \dot{z}_2(t) = -\frac{g}{l}z_1(t) = -\frac{g}{l}\,\varphi.$$

Algorithmic details:

```
% Equation of motion: phi'' + g/l phi = 0
% gravity g, length of the pendulum l
g = 10; l = 1;
% number of time increments, time increment delta_t
n = 100; delta_t = (4.0-0.0)/n;
% initial conditions: angle, angle-velocity
z1(1) = 0.1; z2(1) = 0.0;
% recurrence formula
for k = 1:n
    z1(k + 1) = z1(k) + z2(k)*delta_t;
    z2(k + 1) = z2(k) - g/l*z1(k)*delta_t;
end
```

Fig. 7.17 depicts the (φ, t)-diagram in the time interval $[t_s; t_e] = [0; 4\text{s}]$.

The simulation was performed using $n = 100, 800, 5000$ time increments. For $n = 100$ we have a divergent behaviour in contrast to the simulations performed with $n = 800$ and $n = 5000$ increments.

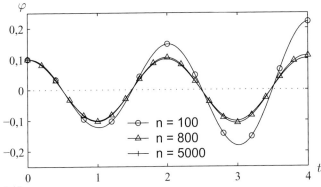

Fig. 7.17

Case ii): Non-linear differential equation:

$$\ddot{\varphi} + \frac{g}{l}\sin\varphi = 0.$$

Transformation to a system of first-order differential equations:

$$\dot{z}_1(t) = z_2(t) \quad \text{and} \quad \dot{z}_2(t) = -\frac{g}{l}\sin z_1(t) = -\frac{g}{l}\sin\varphi.$$

The phase-diagram is depicted in Fig 7.18. For $n = 70$ time increments (crosses) we observe a (incorrect) dissipative behaviour: the time steps are too large and produce an inaccurate approximation. For $n = 3000$ increments (solid line) we observe a closed trajectory in the phase-diagram, which is correct since we have no dissipation.

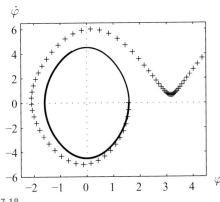

Fig. 7.18

E7.5

Example 7.5 The damped system in Fig. 7.19 is characterized by the parameters $m = 0.5$ kg; $d = 10$ kg/s; $k = 500$ N/m .

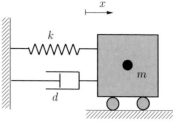

Compute the free vibrations of the system with Euler's method using the initial conditions $x(0) = 0.01$ m and $\dot{x}(0) = 0$. Compare the result with the analytical solution.

Fig. 7.19

Results: see (**C**)   Differential equation:

$$m\ddot{x} + d\dot{x} + kx = 0 \quad \rightarrow \quad \ddot{x} + \frac{d}{m}\dot{x} + \frac{k}{m}x = 0 .$$

Analytical solution (coefficients $\xi = d/(2m) = 10$ s^{-1}, $\omega = \sqrt{k/m} \approx 31.62$ s^{-1}, $\zeta = \xi/\omega \approx 0.032 < 1$, $\omega_d = \omega\sqrt{1-\zeta^2} \approx 31.61$ s^{-1}):

$$x(t) = e^{-\xi t}(A\cos(\omega_d t) + B\sin(\omega_d t))$$

with $A = 0.01$ m and $B = 3.2 \cdot 10^{-4}$ m . Transformation with $z_1(t) = x(t)$ and $z_2(t) = \dot{x}(t)$ to the system

$$\dot{z}_1(t) = z_2(t) \quad \text{and} \quad \dot{z}_2(t) = -\frac{d}{m}z_2(t) - \frac{k}{m}z_1(t) .$$

Fig. 7.20 depicts the numerical solutions in the time interval $[t_a; t_e] = [0; 0.6\text{s}]$ for $n = 50$ and $n = 100$ increments as well as the analytical solution.

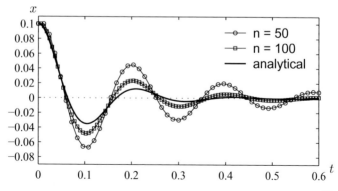

Fig. 7.20

Example 7.6 A point mass ($m = 1$ kg) is subjected to free fall conditions under gravity ($g \approx 10$ m/s^2). The air resistance is assumed to be $A = c\,\dot{x}^2$ with $c = 0.01$ N/m.

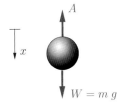

E7.6

Compute the velocity-time diagram with Euler's method and compare the result with the analytical solution.

Fig. 7.21

Results: see (**C**)

Differential equation: $\ddot{x} = g - \dfrac{c}{m}\,\dot{x}^2$.

Analytical solutions:

$$x(t) = \frac{m}{c}\,\ln\left[\cosh\left(\frac{g\,t}{\sqrt{mg/c}}\right)\right] \,,\quad \dot{x}(t) = \sqrt{\frac{mg}{c}}\,\tanh\left(\frac{g\,t}{\sqrt{mg/c}}\right)\,.$$

Transformation with $z_1(t) = x(t)$ and $z_2(t) = \dot{x}(t)$ yields

$$\dot{z}_1(t) = z_2(t) \quad \text{und} \quad \dot{z}_2(t) = g - \frac{c}{m}\,z_1^2(t)\,.$$

Algorithmic treatment:

```
% Free fall with air resistance
% coefficient of air resistance, mass, gravity
c = 0.01; m = 1; g = 10;
% initial conditions
z1(1) = 0; z2(1) = 0;
% number of time increments, time increment delta_t
n = 18; delta_t = (20 - 0)/n;
% numerical integration
for k = 1:n
    z1(k + 1) = z1(k) + z2(k)*delta_t;
    z2(k + 1) = z2(k) + (g - c/m*z2(k)^2)*delta_t;
end
```

In Fig. 7.22 we compare the numerical results for $n = 8$ and $n = 18$ time increments with the analytical solution in the time interval $t = [0; 20\text{s}]$.

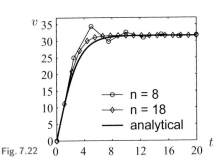

Fig. 7.22

E7.7

Example 7.7   The system in Fig. 7.23 consists of a point mass ($m = 100$ kg;) and an elastic spring ($k = 500$ N/m).

Compute a numerical solution of the equation of motion with the fourth-order Runge-Kutta method for different time increments within the time interval $[t_a; t_e] = [0; 30s]$. Program a short algorithm for the velocity and position updates and compare the results with the analytical solution.

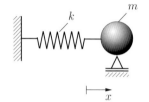

Fig. 7.23

Results: see (**C**)   For the initial conditions $x(t = 0) = 0.01$ m and $\dot{x}(t = 0) = 0$ m/s the diagrams below show the $x(t)$-diagrams for $n = 50$ and $n = 500$ time increments. In the first case we observe a strong numerical (unphysical) damping, whereas for $n = 500$ we obtain an accurate approximation.

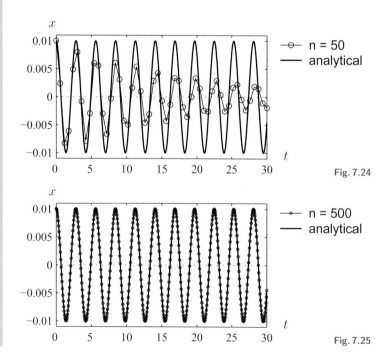

Fig. 7.24

Fig. 7.25

Algorithmic implementation:

```
% Fourth-order Runge-Kutta method
% parameters: mass, elastic spring stiffness
m = 100; k = 500;
% time increment
ts = 0; te = 30; n1 = 50; delta_t = (te - ts)/n1;
% Functions of the set of 1st-order differential equations
f1 = 'z2'; f2 = '-c/m*z1';
% initial conditions
z(1,1) = 0.01; z(2,1) = 0; t(1) = ts;
% numerical integration
for i = 1:n1
    % 1. evaluation
    z1 = z(1,i);
    z2 = z(2,i);
    kk(1,1) = eval(f1);
    kk(2,1) = eval(f2);
    % 2. evaluation
    z1 = z(1,i)+delta_t/2*kk(1,1);
    z2 = z(2,i)+delta_t/2*kk(2,1);
    kk(1,2) = eval(f1);
    kk(2,2) = eval(f2);
    % 3. evaluation
    z1 = z(1,i)+delta_t/2*kk(1,2);
    z2 = z(2,i)+delta_t/2*kk(2,2);
    kk(1,3) = eval(f1);
    kk(2,3) = eval(f2);
    % 4. evaluation
    z1 = z(1,i)+delta_t*kk(1,3);
    z2 = z(2,i)+delta_t*kk(2,3);
    kk(1,4) = eval(f1);
    kk(2,4) = eval(f2);
    % approximate solutions at time t_i+1
    z(1,i+1) = z(1,i)+1/6*(kk(1,1)
                +2*kk(1,2)+2*kk(1,3)+kk(1,4))*delta_t;
    z(2,i+1) = z(2,i)+1/6*(kk(2,1)
                +2*kk(2,2)+2*kk(2,3)+kk(2,4))*delta_t;
    t(i+1) = t(i)+delta_t;
end
```

7.5 Summary

- Equations of motion with initial conditions lead to initial-value problems of first or second order. A differential equation of second order can be transformed to a system of differential equations of first order.

- Most numerical integration algorithms are based on the computation of approximate solutions for differential equations of first order:

$$\dot{x} = f[t, x(t)] \,.$$

- The basic idea of the numerical treatment is the approximation of the time-continuous variation of the function of interest $x(t)$ at discrete points (mesh points):

$$t_i = t_0 + i \,\triangle t \,, \quad i = 0, 1, \ldots, n \,.$$

- Starting from the solution x_i at time t_i (beginning of the time increment $\triangle t$) we compute the (approximate) solution x_{i+1} at time t_{i+1} (end of the time increment) following a specific integration procedure.

- The simplest procedure is Euler's method:

$$x_{i+1} = x_i + f[t_i, x_i] \,\triangle t \,.$$

  In order to achieve an accurate approximation this method may require a large number of time steps.

- More accurate approximations can be obtained by applying the second-order or fourth-order Runge-Kutta method. These procedures require function evaluations at mesh points inside the time increment.

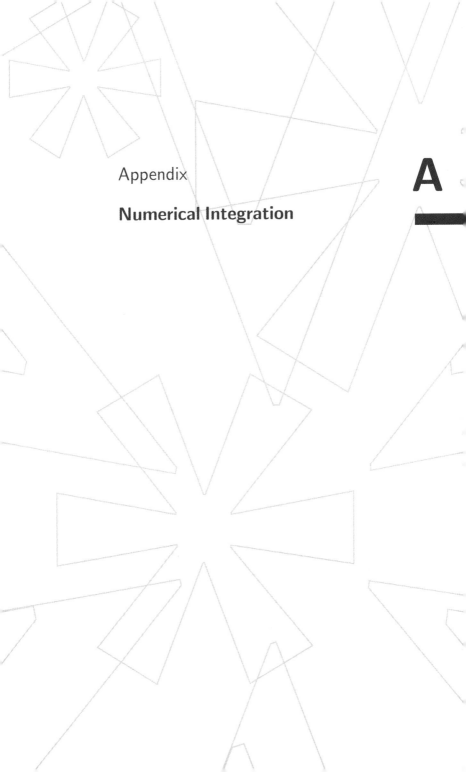

Appendix

Numerical Integration

A

A Numerical Integration

A differential equation of order n

$$y^{(n)}(t) = f[t, y(t), y'(t), ..., y^{(n-1)}(t)] \tag{A.1}$$

can be transformed with the help of the functions

$$\begin{aligned} z_1(t) &= y(t) \ , \\ z_2(t) &= y'(t) \ , \\ ... &= ... \ , \\ z_n(t) &= y^{(n-1)}(t) \end{aligned}$$

to a system of n differential equations of first order:

$$\begin{pmatrix} z_1' \\ z_2' \\ ... \\ z_n' \end{pmatrix} = \begin{pmatrix} z_2 \\ z_3 \\ ... \\ f[t, z_1, z_2, ..., z_n] \end{pmatrix} \quad \rightarrow \quad \boldsymbol{z}' = \boldsymbol{f}[t, z_1, z_2, ..., z_n]. \tag{A.2}$$

To solve this system in the time interval $[t_s, t_e]$ we need the *initial conditions*

$$\{y(t_s), y'(t_s), ..., y^{(n-1)}(t_s)\} \quad \rightarrow \quad \{z_{1_0}, z_{2_0}, ..., z_{(n-1)_0}\} \ ,$$

which are summarized in the column vector $\boldsymbol{z}_0$.

given: $\boldsymbol{z}_0, t_s, t_e, n, \boldsymbol{f}[t, \boldsymbol{z}]$

time increment: $\Delta t = (t_e - t_s)/n$

loop over all time increments

FOR i FROM 0 TO n DO

$\quad t_i = t_s + i \, \Delta t$

$\quad \boldsymbol{z}_{i+1} = \boldsymbol{z}_i + \boldsymbol{f}[t_i, \boldsymbol{z}_i] \, \Delta t$

END DO

Fig. A.1 Euler's method

The numerical solution of (A.2) can be obtained with Euler's method. Its algorithmic procedure is summarized in Fig. A.1.

For the numerical solution of (A.2) with the fourth-order Runge-Kutta method, we have to perform four function evaluations for each of the n first-order differential equations. We assemble these additional function evaluations k_{jl} for $l = 1, 2, 3, 4$ within the column vectors

$$\boldsymbol{k}_1 = \begin{pmatrix} k_{11} \\ k_{21} \\ \dots \\ k_{n1} \end{pmatrix}, \quad \boldsymbol{k}_2 = \begin{pmatrix} k_{12} \\ k_{22} \\ \dots \\ k_{n2} \end{pmatrix}, \quad \boldsymbol{k}_3 = \begin{pmatrix} k_{13} \\ k_{23} \\ \dots \\ k_{n3} \end{pmatrix}, \quad \boldsymbol{k}_4 = \begin{pmatrix} k_{14} \\ k_{24} \\ \dots \\ k_{n4} \end{pmatrix}.$$

A compact program flow of this method is given in Fig. A.2.

> given: $\boldsymbol{z}_0, t_s, t_e, n, \boldsymbol{f}[t, \boldsymbol{z}]$
>
> time increment: $\triangle t = (t_e - t_s)/n$
>
> loop over all time increments
>
> FOR i FROM 0 TO n DO
>
> $\quad t_i = t_s + i \, \triangle t$
>
> $\quad \boldsymbol{k}_1 = \boldsymbol{f}[t_i, \boldsymbol{z}_i]$
>
> $\quad \boldsymbol{k}_2 = \boldsymbol{f}[t_i + \dfrac{\triangle t}{2}, \boldsymbol{z}_i + \boldsymbol{k}_1 \, \dfrac{\triangle t}{2}]$
>
> $\quad \boldsymbol{k}_3 = \boldsymbol{f}[t_i + \dfrac{\triangle t}{2}, \boldsymbol{z}_i + \boldsymbol{k}_2 \, \dfrac{\triangle t}{2}]$
>
> $\quad \boldsymbol{k}_4 = \boldsymbol{f}[t_i + \triangle t, \boldsymbol{z}_i + \boldsymbol{k}_3 \, \triangle t]$
>
> $\quad \boldsymbol{z}_{i+1} = \boldsymbol{z}_i + \dfrac{\triangle t}{6}(\boldsymbol{k}_1 + 2 \, \boldsymbol{k}_2 + 2 \, \boldsymbol{k}_3 + \boldsymbol{k}_4)$
>
> END DO

Fig. A.2 Fouth-order Runge-Kutta method

Index